# Globalizing Organic

# Globalizing Organic

*Nationalism, Neoliberalism,*
*and Alternative Food in Israel*

RAFI GROSGLIK

Published by State University of New York Press, Albany

For information, contact State University of New York Press, Albany, NY
www.sunypress.edu

### Library of Congress Cataloging-in-Publication Data

Name: Grosglik, Rafi, author.
Title: Globalizing organic : nationalism, neoliberalism, and alternative food
  in Israel / Rafi Grosglik.
Description: Albany : State University of New York Press, [2021] | Includes
  bibliographical references and index.
Identifiers: LCCN 2020024656 | ISBN 9781438481555 (hardcover : alk. paper) |
  ISBN 9781438481579 (ebook)
Subjects: LCSH: Organic farming—Social aspects—Israel. | Organic farming—
  Political aspects—Israel. | Natural foods—Social aspects—Israel. |
  Natural foods—Political aspects—Israel.
Classification: LCC S605.5 .G76 2021 | DDC 631.5/84—dc23
LC record available at https://lccn.loc.gov/2020024656

10 9 8 7 6 5 4 3 2 1

# Contents

# Illustrations

# Abbreviations

| | |
|---|---|
| CSA | Community-Supported Agriculture |
| IBOAA | Israel Bio-Organic Agriculture Association |
| IFOAM | International Federation of Organic Agriculture Movements |
| MARD | Ministry of Agriculture and Rural Development (Israel) |
| NOS | National Organic Standards (USA) |
| NOSP | National Organic Standards Program (USA) |
| OFCA | Organic Food Consumers Association (Israel) |
| PPIS | Plant Protection and Inspection Services (Israel) |
| USDA | United States Department of Agriculture |

# Acknowledgments

Writing acknowledgments is the most enjoyable part in the book compilation process. It takes the author back through the sequence of encouragements, guidance, and generosity of so many interviewees, interlocutors, family members, friends, teachers, and colleagues. In the years since I started working on the research for this book, I have lived in three different cities in two continents and in two countries: Israel, a semi-periphery country according to the world-systems theory, and the United States, which is considered a major core country in this theory. The tangible and intellectual crossing—back and forth between Israel and the United States—is a transition that many of my interviewees have experienced. Therefore, the movement between Israel and the Global North (the term I use in this book to refer to the world's developed countries of Europe and North America that are characterized by dominance of world trade, politics and cultural influence) is central both as a conceptual framework of this book and as my own personal experience as the author. From a personal point of view, I have been fortunate to receive support and inspiration from many people in both countries.

First of all, I would like to thank all of the generous people who are engaged in the field of organic food in Israel who welcomed me into their homes, farms, restaurants, offices, and shops, all those who were willing to meet me, talked about their lives, and let me join their activities. Each of them shared with me their thoughts, ideas, experiences, and sometimes the fruits of their labor. They all nourished this project. I hope that they will find this book—including the parts that are critical to their perspectives or those that expose consequences that they did not aim for—interesting and a fair description of their ideas, actions and efforts.

The research for this book started in Ben-Gurion University of the Negev, Israel. I offer my deepest gratitude to Uri Ram, who sparked my interest in social theory in my undergraduate studies and later became my mentor. Uri taught me why both critical theory and robust empirical data matter, provided generous feedback, supported and advanced every step in my development as a scholar, and encouraged me to carve my own path in my research and in academic activities. I am deeply indebted to him for all of these and for his being a source of wisdom and inspiration. André Levy was the first to convince me to rework my project into a book manuscript, which will (hopefully) appeal to an international audience. I thank him, and Ricky Shufer-Levy, for pushing me to write this book, and for their good advice and friendship.

An extraordinary group of scholars at Ben-Gurion University provided me with important insights and helped me clarify my thoughts and ideas in significant ways. I especially thank Nir Avieli and Julia Lerner, who provided intellectual conversations and substantial feedback. They have been incredible collaborators and friends and continue to amaze me with their profound care and support. Special thanks to Daniel Maman for his advice and encouragement. Many other colleagues and friends shared their support and wisdom with me throughout my time in Be'er Sheva. Oleg Komlik, Tama Halfin, Tamir Erez, Yoel Tawil, Noa Leuchter, Einat Zamwel, Adit Shavit, Jonathan Preminger, Yeela Lahav-Raz, Hila Zaban, Dana Kaplan, Dafna Hirsch, Hizky Shoham, Uri Shwed, Sara Helman, Lev Grinberg, Relli Shechter, Eli Avraham, Oranit Klein-Shagrir, Ofra Goldstein-Gidoni, Tally Katz-Gerro, and Liora Gvion provided useful ideas and support in different stages of this project. Tally and Liora have been particularly supportive at several points of this long road and I would like to thank them both. I also thank Dan Kotliar, Tamar Kaneh-Shalit, Ariel Handel, Yofi Tirosh, Aeyal Gross, Alon Levkowitz, and Nurit Buchweitz.

I was fortunate to join Tel Aviv University as a postdoctoral fellow. I am grateful to Dan Rabinowitz for mentoring me at Tel Aviv, for being a source of inspiration and for his support. I also thank Yifat Gutman and Guy Abutbul-Selinger for their collegiality, feedback, and good ideas. At Tel Aviv University I had the great privilege of participating in the Lab for Society and the Environment. I thank the Lab members for their feedback and conversations and for reading parts of my work. I especially thank Talia Fried, Tamar Novick, Nir Barak, Noam Zaradez, and Tamar

Neugarten. Two of my lab friends, Natalia Gutkowski and Liron Shani, have been fantastic colleagues, collaborators, and dear friends. I thank them both for allowing me to share with them thoughts and drafts, for their eye-opening comments, and for the meetings and support during our time together in the Boston area during the writing process of this book, as well as for their friendship over the years.

I have presented portions of this work to various audiences whose comments, suggestions, and questions helped shape my research and arguments. Among the many organizers, discussants, and engaged audiences, I am especially grateful to Alan Warde, Bente Halkier, Josée Johnston, Motti Regev, Gisela Welz, Yahil Zaban, Ronald Ranta, Wendy Wills, Carole Counihan, Jennifer Smith Maguire, Penny Van Esterik, Rachel Soper, Oz Frankel, Nahum Karlinsky, and Daniel Welch. Some of my research was published in Hebrew (2017). I would like to thank Idan Zivoni and Itzhak Benyamini, editors at Resling press, for being the first to believe that my research is worthy of publication as a book. I also thank Garrett Broad and Daniel Monterescu for their assistance in preparing the book proposal for this manuscript.

I am grateful to have spent time at the Brandeis University's department of Sociology in 2015 to 2017. I have been blessed with the support and generosity of Laura J. Miller, who became a spectacular mentor, a source of inspiration, and a friend during my two years at Brandeis. I am grateful for the multiple conversations, invaluable feedback, and many kinds of support she has given me. I owe her a great personal and intellectual debt. I am also much indebted to the many friends and colleagues whom I met during my time in the Boston area, including Merry White, Ari Ariel, Andrea Borghini, Steven Shapin, Barbara Haber, Gus Rancatore, Ben Wurgaft, Shai Dromi, and Darra Goldstein. I thank them for the engagements with my work and for sharing with me their thoughts and ideas and in some cases their wonderful food, coffee and wine. Graduate students in my seminar on ethical eating at Boston University had a considerable impact on this book. Their questions and interest in my project helped me think deeply about how to articulate my arguments and how to make them accessible.

My time at University of California, Davis has been significantly important in the final and significant stages of preparing this manuscript. I owe a huge debt to Diane Wolf for paving many ways for me in the three exciting years since I joined the Department of Sociology as a visiting assistant professor. I would like to thank Diane, from the bottom of my heart, for providing me with multiple resources for advancing my research and completing the manuscript for this book, as well as for her encouragement and support throughout these years. I am also profoundly grateful to David Biale for his assistance and advice. I thank many of the faculty, administration, staff and students at UC Davis for developing an extraordinary friendly, supportive, and inspiring environment. Special thanks go to Ryken Grattet, David Kyle, Laura Grindstaff, and Bruce Haynes. I would also like to thank Susan Gilson Miller, Eva Mroczek, Sven-Erik Rose, and Zeev Maoz for helping me feel at home in the Jewish Studies program.

The Israel Institute provided crucial funding for my fellowship and in support of this project. I have benefitted from the insightful feedback from graduate students in my seminars on food movements and social justice and on sociology of consumption in UC Davis. I am grateful to Charlotte Glennie, Alana Stein, Nadia Smiecinska, Morganne Blais-Mcpherson, Gwyneth Manser, and Sasha Pesci, who read the near-completed manuscript or individual chapters. They all went above and beyond, providing substantial comments, edits, and suggestions that helped make this book better. I wish to thank all the participants of the UC Davis Food Systems discussion group for their feedback. I am also grateful for insightful and productive exchanges with Alison Hope Alkon, Julie Guthman, Charlotte Biltekoff, Catherine Brinkley, and Ryan Galt.

Altogether, I have profound gratitude for everyone in California for making Davis a wonderful place to have completed this book.

I thank Dakota Mattson for her edits and comments on English language articulations and Michele Tobias for her assistance with my mapping needs. Very special thanks are owed to Rafael Chaiken, my editor at SUNY Press, who shared his enthusiasm for the project and for his steady hand in guiding the process of publication. I also thank Eileen Nizer and Anne Valentine. I am grateful to the two anonymous reviewers whose comments provided helpful suggestions for refining the manuscript.

I owe my deepest gratitude to Yehudit and Oded Shoham, who, more than anyone else, worked hard and supported me in so many ways. This book would not have been written without their efforts, generosity and tremendous help. Thanks also to my parents and brothers Alon, Idan, Liora, and Zeev Grosglik for their care.

I cannot thank my loved ones enough—Einat, Michal, and Dror. Trying to describe how much I thank them for being there with me for all these years is where words could never do justice.

*Source:* OpenStreetMap (green line), US Department of State (countries), and Natural Earth Data (lakes). Cartography: Michele M. Tobias.

# Introduction

## Organic Hummus

"Is there anything to say about hummus that has not been said so far?" was the title of a newspaper article by Ronit Vered, a food writer for *Haaretz* (an Israeli newspaper), which summarized a series of symposia on "hummus" held in Haifa and Jerusalem.[1] Indeed, in recent years it seems impossible to discuss anything about food in Israel without mentioning the Arabic word "hummus." I have yet to meet one academic scholar, food writer, celebrity-chef, or restaurateur who did not mention hummus while discussing food in Israel.[2] I will not be the exception either, and therefore I must admit: hummus is an intriguing and polysemic food. Although it is a relatively simple and mundane dish, it is loaded with many social meanings. Hummus is now a key culinary symbol[3] of the Middle East and of Mediterranean cuisines. It signifies rootedness, earthiness, local simplicity, authenticity, commensality, and *baladi* (an Arabic term often used by Jews and Arabs in Israel that means "rural," "rustic," or "folkish"). It is also a food that implies culinary cultural appropriation, culinary colonialism, and nationalism.

On January 8, 2010, I attended an event in the Arab-Palestinian village Abu-Ghosh in Israel: an attempt to break the Guinness record for the biggest plate of hummus in the world. The master of ceremonies and the honored speakers proclaimed, "Breaking the record is of national pride; the Lebanese claim that hummus is their invention, but hummus is ours, the Israelis.'" Hundreds of people cheered and gathered around a huge plate holding four tons of hummus that had been prepared by two local industrial food companies. The entire event, which was one battle in what was named "the hummus war"[4] between Israel and

1

Lebanon, received impressive attention in the public sphere and even among scholars. It was interpreted as a case study of varying sociological and cultural processes termed gastro-politics, gastro-diplomacy, and gastro-nationalism.[5]

Surprisingly, it was precisely in such a local, politically loaded event that I discovered that hummus had also become imbued with environmental and health meanings. Esther, a clinical dietician whom I met at the event, disputed the excitement around us:

> Does it matter if hummus is ours or belongs to the Lebanese? The hummus here isn't even real hummus; it is just an industrial product, a spread, just like a store-bought mayonnaise. Real hummus is made by hand, from real chickpeas without artificial additives and preservatives as in this stuff. I love hummus, but not the industrial kind. The healthiest and best hummus is made from *organic chickpeas* and from whole *organic tahini*. Besides, what will they do with all this hummus? It will be thrown out. As if our environment isn't polluted enough.

It was the first time I heard about organic tahini (sesame paste) and organic chickpeas. At that time, I had just completed a research project on processes of globalization, McDonaldization, and the social construction of authenticity as seen through the lens of food in Israel.[6] When I was first introduced to the concept, and the dish, "organic hummus," I was intrigued. How have progressive meanings related to organic food—such as sustainability, and resistance to global-industrialized food production—become tangled with local, political, and national symbols such as those embedded in hummus? Later, after tracing "the social life"[7] of organic hummus, I realized that some of the mechanisms that allow the existence of such a dish underlie the questions addressed in this book: How, and to what extent, are global ideas associated with the notion "organic" and entwined with local culinary and agricultural contexts? How was organic agriculture integrated in Israel—a state in which agriculture was a key mechanism in promoting Jewish nationalism (Zionism) and in time, has been formulated into a highly mechanized and technologically sophisticated field? What are the social, environmental, and economic consequences of the development of organic agriculture in contested places such as Israel, the Gaza Strip, or the West Bank? In what ways is organic food recruited for the construction of social identities and how has it intersected with, or been impacted by, local foodways and lifestyle practices? What practices do producers, consumers, distributers, and food purveyors in Israel use to create symbolic and economic values of the notion "organic"?

Before considering these questions in detail, it may be helpful to take a closer look at the case of organic hummus, which touches on the different conditions in which organic foods are produced, distributed, and consumed in Israel (and beyond), and exemplifies some of the main manifestations of the globalization of organic.

## "There Is Nothing More Organic than Hummus"

The word *hummus* often serves, in Israel, as an abbreviation of the full Arabic name of the dish—*hummus bi tahini*: dried chickpeas soaked in water, boiled until softened, crushed, and mixed with tahini and often served with olive oil and fresh pita bread. Organic hummus consists of the same ingredients, of course, albeit ones that are labeled as organic. The main ingredient of the dish is chickpeas. Usually, the chickpeas used for making hummus in Israel are grown locally. In recent years, farmers began growing organic chickpeas in Israel. Nevertheless, a large part of the chickpeas grown in Israel, including organic chickpeas, are exported. One of the restaurateurs who serves organic hummus told me that when he cannot get locally grown organic chickpeas, he doesn't hesitate to use imported chickpeas. "The question of the chickpeas' origin doesn't really bother my customers. As long as it's labeled organic—they're okay with it."

Unlike chickpeas, tahini—the second most important ingredient in this dish—is definitely not made from local ingredients. Though in the past sesame seeds were grown within the geographic area known as "the Land of Israel"/Palestine, the cultivation of sesame ceased several decades ago. Nowadays, the vast majority of sesame seeds consumed in Israel are imported from countries along the equator—mainly from Ethiopia but also from Guatemala, Uganda, and Eritrea. However, tahini is symbolically considered a "local" product. Part of this symbolism emerges from the fact that the processing and grinding of sesame seeds are done in Israel/Palestine,[8] but mainly because of the use of tahini with falafel (fried ground chickpea patties served in a pita bread pocket)—previously a national culinary symbol in and of itself.[9]

Almost all the venues that serve organic hummus use tahini produced by a factory situated beyond the Green Line[10]—a Palestinian territory under Israeli military control. After talking to numerous organic hummus manufacturers and consumers, I realized that they do not see this as a problem. Their main consideration regarding organic tahini rests with the extent of its "organicness," healthiness, and quality, while neglecting tenets of social justice and fairness that are often ascribed to the notion "organic."[11] But the case of organic tahini in Israel raises an intriguing question: What allows the attachment of the label "organic" to a food item which is made of imported ingredients from a developing country and manufactured under controversial political circumstances?

Questions about the relation between organic food, import, and export are also raised from looking at the third main ingredient in any hummus meal: pita bread. The consumption of hummus in Israel/Palestine is mostly accompanied by pita bread made from refined or bleached flour. However, organic hummus is served with pita bread made from whole wheat flour. The inclusion of whole wheat pita bread seems to indicate the "organicness" of the dish. Whole wheat pitas can be easily recognized by their taste, texture, and brown color. They are perceived as healthy and as locally handmade. Thus, whole wheat pitas are culturally constructed as a local food. However, most of the wheat used to manufacture flour in Israel (including whole wheat organic flour) is imported.

When looking at the distribution of organic hummus, one can find that the marketing processes of this dish entail meanings of "Israeliness," cultural appropriation, naturalness, healthiness, and cosmopolitanism. Take, for example, what seems to be the first restaurant to boast the title "organic" in Israel: Aba Gil (see Figure I.1). This restaurant (which operated in Tel Aviv between 2005 and 2015) was actually an organic hummus restaurant, or *hummusia*—an eatery where artisanal hummus is made and sold. This *hummusia* was described as "the first Zionist organic restaurant"[12] by the local Israeli media. *The Guardian* (a British daily newspaper) and tour guides of Israel mentioned it and recommended trying "the national dish"—not only in its "regular" or "industrialized" version but also the "organic, more authentic version."[13]

The name of the restaurant, Aba Gil, is a "Hebrewization" of the Arabic word *abu* attached to popular Arab hummus joints in Israel (for example, Abu Hassan or Abu Shukri). Both *abu* and *aba* mean "father," in Arabic and Hebrew respectively. Aba Gil served, in addition to organic hummus, foods that were defined by the owner as vegetarian, healthy, and environmentally friendly. But organic hummus was the most popular dish served at Aba Gil. The owner explained: "Hummus has qualities that are fundamental to the Middle East. Hummus speaks

Figure I.1. Aba Gil: Organic hummus. Photo by the author, September 2010.

for itself; it exists. It is *the* food here. There is nothing more organic to this place than hummus." But contrary to the locality he ascribes to hummus, his restaurant had many global-cosmopolitan similes. For example, the decoration of the restaurant resembled tourist restaurants in India or Nepal: low tables and futon sofas, a serene ambience created by soft "world music" or meditation music, and a large bulletin board covered with countless flyers about yoga, holistic healing, and spiritual activities. Combined with organic hummus, the restaurant seemed to be an Israeli version of global New Age culture and the "Far Eastern" countercultural style.[14]

Organic hummus is also prevalent in retail chains that specialize in marketing organic, "green," natural, and healthy food (*superim organim* as these Israeli venues of global ethical and "green" consumerism are called).[15] Inside some of these organic supermarket chains, one can find organic hummus counters. These counters offer organic hummus, whole wheat pitas, organic whole tahini (tahini made from whole sesame seeds), organic falafel, and organic chopped vegetable salads. The counters operate according to organizational principles that sociologist George Ritzer associated with McDonaldization—efficiency, calculability, predictability, and control.[16] The organic hummus is ready-made, the menu is limited to a few items displayed on large backlit posters, and the employees, who wear green uniforms (that symbolize "green" environmentalism) are temporary workers. Furthermore, the organic hummus counters are named a hybrid *glocal*[17]expression: *organic hummus bar*. This glocal name is similar to a burger bar or sushi bar, representing global-Western updated culinary trends (bar) and, allegedly, a local dish (hummus) (see Figure I.2).

At one hummus bar, on the counter where diners ate, there were placemats depicting a reference to hummus in the Bible (see Figure I.3): "Eden Hummus Bar: King David—was born thanks to . . . hummus!!" These placemats include a biblical quote: "And at meal-time Boaz [King David's great-grandfather] said unto Ruth [King David's great-grandmother]: Come hither, and eat some of the bread, and dip thy morsel in the hometz" (Book of Ruth, 2–14).[18] The text goes on to explain that the word *hometz*, which in Modern Hebrew refers to vinegar, actually refers to hummus. The word *hometz* not only sounds like *hummus* but also resembles the word *himtza*, which is the Hebrew botanical name for chickpeas.[19] Clearly,

Figure I.2. Eden Hummus Bar in Tel-Aviv: Fast-food organic hummus. Photo by the author, June 2012.

Figure I.3. The placemat from Eden Teva Market. The title of the text says: "Eden Hummus Bar: King David was born—thanks to . . . hummus!!" Photo by the author, July 2014.

these placemats represent a symbolic connection between the "organic-ness" of the hummus in these hummus bars and the biblical Kingdom of Israel. Therefore, one can see them as a "gastro-political weapon"[20] in the above-mentioned "hummus war," and in the prolonged debate about "who invented (and owns) hummus?"

## Organic for Change: Organic Food and the Sociology of Alternative Food Movements and Ethical Eating

The case of organic hummus shows how the notion "organic," which seems to represent global eco-social ethics, is shaped by local cultural, political, and economic processes.[21] It also mirrors how varied cultural and ideological tendencies, such as environmentalism, neoliberalism, and nationalism, enter the realm of mundane social activities, such as food production, consumption, and eating.[22] Ultimately, the case of organic hummus illustrates how organic foods are essentially complex, polysemic culinary artifacts.

Rooted in the anti-industrial movements of the early twentieth century, organic agriculture and organic food movements are often described as the "remedy" to the hazards of market growth–driven food systems. Since its inception, proponents of organic farming in Europe and North America have advocated for an alternative to scientific (chemical) and industrial agriculture.[23] Agronomically, they were troubled by the depletion of the biological complexity of soils cultivated by modern scientific-chemical methods. Advocates of organic farming were also concerned with "severe consequences such as desertification and the toxicity of agricultural chemicals."[24] They resonated with philosophical thinking about the interconnectedness of nature and the relations between environment and society, humans, and non-humans. Some of these advocates encouraged a romantic nostalgia for agrarianism and a longing for the communal conditions of a lost countryside.[25] From a political-economic perspective, they saw organic farming as a locus of environmental sustainability and as an effective expedient for balancing uncontrolled and unjust market growth. In the United States, for example, the organic discourse intensified throughout the countercultural era of the 1960s and the environmental movement of the 1970s. Organic farmers and consumers played a prominent role in developing and promoting a "counter-cuisine" that challenged conventional systems of food production and distribution.[26] Later, organic agriculture was developed from a fringe of countercultural farmers to an institutionalized niche within the food industry.[27] In France, organic farming has been practiced since the 1940s, but it was not until the 1970s that it flourished, mainly in resistance to the "productivist model" of conventional French agriculture.[28]

Throughout the institutionalization and proliferation of organic farming, the adjective "organic," which in its contemporary scientific meaning stands for a living organism or for a part of a living organism, began to be ascribed to a product of a plant or an animal grown without synthetic pesticides, fertilizers, hormones, antibiotics, or genetic modification, as well as high standards for the treatment of livestock. Nowadays, the term "organic" is widely seen as embodying meanings of naturalness, healthiness, and safety. But it is also often considered not only a food that is "good to eat" but also a virtuous food, a food that is "good to think with" (to paraphrase Claude Lévi-Strauss[29]) and engages with moral social and environmental issues. In fact, organic food is often considered the vanguard of foods labeled as ethically produced.[30] Therefore, organic farms have gradually become charged with meanings and images such as family-owned and local agricultural operations, small-scale and sustainable agrarianism, and spaces where human and environmental health are prioritized. Accordingly, a preference for organic foods by consumers is conceived as an ethical eco-social act, or, to put it differently, as a practice of citizenship and morality.[31] These ethical meanings are manifested within "the principles of organic agriculture" as articulated by the International Federation of Organic Agriculture Movements (IFOAM): health, ecology, care, and—as previously mentioned—fairness.[32]

As the sales of organic food began to rise in the last two decades[33] (mostly in North America, Europe, and, to a lesser extent, Asia and Latin America), the notion "organic" featured in public and scholarly debates concerned its likelihood of making eco-social change. As part of this debate, an important sociological critique toward organic agriculture and organic food consumption has emerged. Critical geographer Julie Guthman, for example, argues that the production and sale of organic (and local) foods does not challenge the corporate food system but merely creates niche markets that function as an insignificant alternative to conventional agriculture.[34] Together, and in parallel, to other scholars, Guthman criticized the rise of "corporate organic" in the Global North and the transition of organic from a countercultural movement to a market-oriented sector.[35] "Big Organic," "Organic Lite," "Organic Industry," and "Organic-Industrial Complex" are all critical notions associated with the processes of industrialization and conventionalization of organic that took place beginning in the early 1990s. Critical scholarly works on organic food in the United States argue that contemporary political and economic processes subverted the socio-ecological vision associated with the origin of organic agriculture.[36] Organic farmers in places such as California were driven to use intensive and industrial techniques, similar to those used by conventional growers.[37] Organic movements in places other than the United States and Western Europe have also been described as co-opted by large-scale food conglomerates and retail chains. As such, the more radical goals of organic movements were put aside, and the ethical meanings ascribed to organic agriculture—those that made organic food an important emblem for resistance against the conventional food sector—were diluted by commercial interests.[38]

Practices of organic food consumption have also been critically examined. It is indicated that practices of organic food consumption in the Global North are associated with collective environmental and social care combined with individualistic motivation such as healthy diets and good taste.[39] Accordingly, organic food consumers (as well as other ethical eaters) are often portrayed as "citizen-consumers," or "reflexive consumers," namely those who seem to think critically about the social and ecological problems of the industrialized food system, propose solutions, and even engage with food through more self-aware and democratic processes.[40] However, empirical studies that examined this hybrid conception of "citizen-consumer" revealed that consumers were attracted to organic products primarily because of the quality and palatability and were often motivated by care for their personal and family members' health.[41] Theoretical and publicly debated concepts, such as "the reflexive consumer" or "the ethical consumer," have been fetishized as a major locus of social change.[42] In this regard, the fact that eco-products are mostly accessible to privileged consumers has been addressed critically alongside the ways in which such accessibility might position organic consumers as thoughtful and sustainability-minded and thus may serve as a means to demonstrate high social status. Existing critiques of organic (and local) eating as "bourgeois piggery" or as "yuppie chow" point out that the ideology of consumerism often facilitates the link between political eating and status, thereby fostering troubling politics of gender, class, and race.[43]

As part of broader accounts on current modes of alternative foods movements and alternative foods networks, both organic food production and consumption have been critiqued for embodying a neoliberal worldview and market-based logics.[44] The neoliberal-consumer-ethic frame suggests that individual entrepreneurialism can solve collective environmental and social problems, and that consumer choice is a primary pathway to social change. Accordingly, neoliberal political economies and subjectivities focus mainly on engagement with systems of voluntary regulation rather than on trying to leverage state and corporate power to pursue change.[45] Therefore, critical perspectives on the organic conceive it to be a liminal notion blurring the boundaries between (1) markets and social movements, (2) lifestyles and activism, (3) agricultural development and romantic agrarianism, (4) commodities and ideologies, and (5) nature and culture. Thus, it has been claimed that there is some "gray area" between the "alternative" and the "conventional," and that practices supposed to be counted as "resistance" and "oppositional" are, in many respects, "analog to the very things they are purported to resist."[46]

This book corroborates some of the accounts mentioned above. However, these claims draw mostly on case studies in the US, Canada, and Western Europe, whereas organic agriculture and organic food consumption outside the Global North is an understudied area of research.[47] Therefore, the analysis of the emergence and development of organic food in Israel presented here intends to expand this scholarship by emphasizing a somewhat overlooked aspect embedded in the alternative food movement: the tension between the global and the local. As discussed below, foods labeled as ethically produced, including organic products, are transformed into global culinary artifacts in and of themselves. While the existing literature mostly addresses how organic movements were transformed over *time*, this research looks at how the notion organic changed in relation to *space*. Put differently, the ways in which ideas about "organic" have been implemented as they moved from the Global North across societies, cultures, and political borders have been neglected.

Grounded in the studies of the globalization of food—which look at the ways in which cross-political and cross-cultural processes affect the realm of food, and how food-related processes affect the globalization of the social order[48]—this book proposes a different view for understanding the social complexities of organic; focuses on the interactions, tensions, and limitations of the meeting points between global and local processes; and uncovers the ramifications of these processes in the field of alternative foods.

## Organic (and Other Alternative) Foods as Global Cultural Artifacts

So far, I have discussed organic food (and other similar food-related categories such as slow food, local food, fair trade food) as an alternative to the destructive and exploitative consequences of the industrialization of food. Organic agriculture and organic movements are commonly conceived as a prominent trend that seeks to reverse the "gastro-anomie"[49] that, arguably, characterizes modern practices of food production and consumption and to challenge the profit-driven global agro-food system.[50]

Symbolically, the proliferation of organic food (and local food, fair trade, slow food, and the like) seems to be antithetical to the McDonaldization of food systems, namely the incorporation, standard-ization, globalization, and degradation of agriculture as well as to the global homogenization of eating culture.[51] Giovanni Orlando explains the emergence of organic farming and organic food consumption as a symbolic expression of reflexive modernization and an expression of the development of a "risk society," namely recognition of the negative impacts on nature by society, of the difficulties of maintaining scientific dominance over nature, and of the consequential rise of risks due to human activities.[52] According to Orlando's account,[53] organic food (as a substance and as a symbol) is oppositional to the "icons of risk,"[54] such as genetic modification, chemical fertilizers, $CO_2$ emissions, and the loss of biodiversity, which are all closely related to the globalization and industrialization of food systems. Continuing this line of thought, and referring to the conceptualization of globalization as a process of "disembeddedness" or a change in the scale of human interconnectedness and a "lift out" of all social interactions from local contexts,[55] organic and local foods can be conceived as culinary artifacts that work toward re-embedding the relationships between society and technology, body and environment, nature and culture, and food and humanity. From a normative perspective, organic agriculture represents the creation of an "alternative food system" or even a realization of processes of "alter-glo-balization," namely efforts intended to shift existing exploitative processes of globalization to alternative and virtuous processes of globalization.[56]

Yet, despite the widespread discourse on the "localness" or the "alter-global" traits of organic, it is noteworthy that organic farming has been globalized structurally and institutionally. Historian Gregory Barton mentions that following the (relative) acceptance of the organic narrative in Britain and the United States after World War II, the organic movement made startling progress in terms of the influence and growth of consumer markets.[57] During the 1970s, organic farming was formed into an international movement. The International Federation of Organic Agriculture Movements (IFOAM) was founded in 1972 in Versailles, France, by a group of farmers from Britain, France, Sweden, South Africa, and the US engaged in organic and biodynamic agricul-ture. By 1975, IFOAM grew and had dozens of members representing 17 countries.[58] Since 1980, IFOAM has been working in partnership with the European Community (EC) to establish organic farming certification and integrate organic farming practices into national and international government policies, certification standards, and legal definitions.[59] At that time, in the United States, states such as Oregon, Maine, and California adopted policies and legal definitions for the use of the term *organic*. In 1990, the National Organic Standards Program (NOSP) was established. The NOSP required the United States Department of Agriculture (USDA) to implement uniform national standards for organic goods (National Organic Standards [NOS]) as well as to guide and regulate organic farming. Concurrently in Europe, several countries institutionalized their own organic certification procedures. Later, IFOAM's standards for certification led the European Union to adopt an EU-wide organic certification program. In 1991, EC regulation 2092/91 on organic production of agricultural products was published by the European Commission; it was enforced as law in 1993. In other countries, such as Canada, Australia, New Zealand, South Africa, and Japan, organic farming practices developed parallel to the development of organic farming in Europe and the United States.[60]

The trade agreements on organic food at the international level that were signed in the 1990s and the 2000s created a new global robust circulation of organic foods.[61] In 2016, organic products, with a total value of almost US$90 billion, were sold globally.[62] Between 2000 and 2010, the organic market more than tripled (2000: US$17.9 billion; 2010: US$59.1 billion).[63] Farmland for growing organic produce increased globally from 11 million hectares managed by 200,000 producers in 1999 to 57.8 million hectares managed by over 2.7 million producers in 2016.[64] Over the first decade of the twenty-first century, China, Korea, Africa, and Latin America became large producers of organic agricultural products due to the globalization of certification standards and export centered economies.[65] The increase in the volume of production and trade led to the expansion of the range of organic food products available to consumers around the world, mostly in the Global North.

Currently, about 90% of the sales of organic produce take place in North America and Europe. In 2016, for example, the countries with the largest organic markets were the United States, Germany, and France, and the highest per capita consumption was in Switzerland and Denmark.[66] The distribution and marketing of organic food also changed completely from its inception to the present with methods ranging from small stores in agricultural communities, farmers markets, community-supported agriculture ("box schemes" in the UK), and transnational grocery chains labeled as "organic" to online organic food marketing (the latter will probably continue to rise, as one might assume from that fact that the online retailer Amazon purchased the organic supermarket chain Whole Foods Market in June 2017).

The globalization of organic food as this book intends to exemplify, however, is not only about the circulation of a set of farming techniques, commodities labeled organic, or certification and regulation standards across the globe. The ethics, aesthetics, and politics of organic food, as well as the global repertoires of ethical eating, do not operate solely according to technical-agricultural or organizational rules. Rather, the social meanings of organic food have been circulated, translated, and articulated differently—both across and within nation-states. They have been assimilated and implemented according to particular historical, cultural, political, and economic contexts. Furthermore, organic has not always served as a means of resistance to the global-industrial food system, and organic food consumption is not always a matter of ethical consumption.

As I will argue in this book, organic food has evolved as part of the globalization of agriculture and culinary culture, and it is a matter of broad processes of economic and ideological neoliberalization, local-ization, post-nationalization, and neo-nationalization, no less than a matter of resistance or "alternativization."[67] Ideas related to organic may appear as part of the global diffusion of environmental ideas, along the lines of the proliferation of healthy-lifestyle movements, or as a cultural trend of "eating local," and may be realized and loaded with meanings that do not necessarily undermine processes of globalization. They may also be interwoven with local foodways, with routine eating and cook-ing practices,[68] and with certain "histories"[69] of agricultural produce, foods, and dishes. Organic philosophy and practices can be intertwined with local similes attributed to agriculture and with the various local political meanings associated with agrarianism. They may also intersect with local theological cultures, with local political ideologies (such as nationalism in the case of organic hummus), with contested symbolic social relations, and with claims of power, control, or position in the sociocultural hierarchy (such as colonialism and cosmopolitanism in the aforementioned examples of organic hummus). As the case of the emergence of the organic agricultural sector and the development of a field of organic food in Israel/Palestine exemplifies, the virtuous mean-ings attributed to practices of organic food production and consumption are always in circulation, and these, once mobilized, are used by local producers and consumers to negotiate particular issues of taste, morality, and identity. Apparently, there is no rigid organic philosophy or a set of organic procedures that can be applied straightforwardly without cultural, institutional, and technical translation, or without adaptations to local conditions.

## Organic Food in Israel: Introduction

The Israeli field of organic food is an exemplary case of the globalization of alternative food movements in the twenty-first century. But it is also a significant case in and of itself, because of the centrality of agriculture in both the Israeli-Zionist and the native Palestinian ethos and practices. Since the first decades of the Zionist project, Jewish (conventional) agriculture in Israel was a key mechanism that promoted Jewish nationalism (Zionism).

The Israel Bio-Organic Agriculture Association (IBOAA) was established by a group of Jewish farmers at the beginning of the 1980s. They were the first to introduce organic agriculture to Israel. As I will show in Chapter 1, their rhetoric—which criticized the extensive use of chemical fertilizers and pesticides in Israel's conventional agriculture—was entwined with national ideology, myths, and symbols. Similar to conventional Jewish agriculture in Israel, they combined these ideological meanings with pragmatic, rational-instrumental practices of efficiency, constant growth, and aspiration to become leading organic exporters in global agricultural markets. Currently, the IBOAA serves to unionize approximately 600 farmers and food producers. In 2017, 693 people, about a half of them farmers, were registered and worked in organic food production (compared to 640 in 2016).[70] From 2005 to the present, approximately 17,000 to 20,000 acres of land in Israel have been cultivated by organic farming methods. These areas constitute between 1.7% and 2.7% of the total area used for agriculture in Israel. In 2017, the total area of land that was cultivated by organic farming methods grew by 20% compared to 2016.[71] Currently, crops cultivated by organic methods account for about 1.5% of all agricultural produce in Israel.[72]

Organic agriculture and organic food production in Israel are export-oriented. Nowadays, about 92% of organic produce is exported, and the rest is traded within the local market. Organic agricultural products account for between 8% and 13% of the total export of fresh produce from Israel.[73] The rate of growth in Israeli organic agricultural produce is reflected in export data. In 2006, Israeli organic food exports amounted to 45,000 tons of fresh agricultural produce. In 2008, export of organic agricultural produce from Israel almost doubled, amounting to 80,000 tons. Between 2011 to 2012, exports of organic agricultural produce to EU countries alone amounted to about 67,000 tons, and by 2013, nearly 90,000 tons of organic fruits and vegetables were exported to Europe. Organic produce, with a total value of ₪1 billion (1 billion new Israeli shekels), was sold between the years 2008 to 2012 (about US$270.5 million). In recent years, there has been a growing demand for organic food products by the local Israeli market. While in the year 2000, total sales of the organic food market within Israel amounted to between ₪15 and ₪25 million, by 2010, total sales amounted to ₪300–400 million.[74]

Moreover, the sale of organic food products in 2005 amounted to approximately 0.35% of the total food consumption, whereas in 2010 the total sale of organic food products in Israel increased to around 0.6% of the total food consumption.[75] In 2005, it was estimated that 0.7% of the population in Israel consumed organic food regularly, and another 1% consumed it once in a while.[76] According to a survey of sustainable consumption in Israel from 2010, 14.5% of households regularly (but not exclusively) consumed organic food, 13.5% frequently consumed organic food, and 23% rarely consumed organic food.[77]

The increase in organic food sales can be understood as part of broad changes taking place in Israeli eating habits and food consumption. During the pre-State period (1882–1947), and for a decade after the establishment of the State of Israel in 1948, the dominant public discourse among the Jews and the Arabs in Israel treated eating as an act of physical maintenance. Food and agriculture were used as a symbolic and discursive means of cultivating Zionist identity among Jews who immigrated to Palestine.[78] Palestinian agriculture was considered inferior,[79] and many of the eating habits of native Palestinians (whose political and cultural identity have been shaped—since 1948—by civil limitations and institutionalized discrimination) had been erased, overlooked, and in some cases, appropriated.[80]

From the 1950s to the 1980s, the development of Israeli foodways and food production were subject to economic crises, high-intensity processes of migration, structural problems of distributive justice, and conflicts between ethnic and national groups, such as Jews of European origin (Ashkenazim); Jews from North Africa, Central Asia, and Arab countries (known as Mizrahim, Sephardim, or Oriental Jews); and Palestinians. Thus, any attempt to usher in new eating styles or methods of food production (such as growing or consuming organic food) faced the barriers and limitations of complex social conditions.

Beginning in the mid-1980s, Israeli society underwent an accelerated process of economic neoliberalization and the strengthening of a global consumer culture and lifestyle.[81] Consequently, the status of food in Israel rose, and a new Israeli culinary discourse began to flourish. It was reflected in a significant process of the Americanization of food production and consumption as well as in the proliferation of foods representing "ethnic" and "exotic" images.[82] During the first decade of the new millennium, Israeli society was already well embedded in the globalization processes, which paved the way to processes of deregulation and privatization of many sectors of agriculture and food production. Yet Israeli society labored (and still labors) under a volatile regional conflict with limited political and institutional resources devoted to solving environmental and global-ethical problems.[83]

Recently, however, people living in Israel—as in other parts of the Western world[84]—have become increasingly cognizant of the political and cultural implications of their food choices and practices. Several kinds of alternative culinary discourses, voluntary food justice movements, food-related lifestyle movements, and policies for sustainable food production have emerged in Israel. Among them are veganism, meatless Mondays, fair trade products, and the implementation of national policies for sustainable agriculture. Since the mid-2010s, the alternative culinary discourse obtained such a significant foothold among certain groups in Israeli society that Gary Yourofsky—a globally renowned American animal rights activist—declared that Israel had become the mecca of veganism and food activism.[85] Discursively and institutionally, organic food agriculture and culture in Israel was ahead of all of these movements and was formed as the first Israeli alternative foods movement.

Thus, one can ask how organic food, which signifies environmental protection and social equity, has been realized in Israel, a country in which environmental issues are perceived as less pressing than to inner political conflicts or the Israeli–Arab conflict and recurrent wars? Is growing organic vegetables in the Jerusalem area or in the West Bank different from growing organic vegetables in, for example, Northern California? Is consuming organic food in Tel Aviv similar to consuming organic food in Western European, American, or Canadian cities? In this book I attempt to expand the understanding of organic culture and agriculture through focusing on the Israeli case, while at the same time looking through the lens of organic food in order to discuss Israeli cultural, political, and economic currents.

In the chapters that follow, I marshal the ethnography and qualitative data that I collected (see notes on methods, below) to engage with the questions mentioned above in a way that goes beyond focusing on farming practices, organization, or certification processes. I show that organic food production and consumption are not only informed by Western forms of environmental and sustainable agriculture but also shaped by local ideology and political economies. I address the intersections and discrepancies between structural and symbolic levels of organic, both from local and global perspectives, and show how they are realized according to local versions of global neoliberal consumerism, food politics, local foodways, and local media and popular culture.

As an extension of existing literature on the social aspects of organic agriculture and food movements—which tend to tackle the different levels of alternative foods separately (focusing on the level of either production, consumption, distribution, or mediation)—I will focus on organic agriculture and organic culture more broadly. The following chapters offer an integrative analysis of the levels of production, consumption, distribution, politics, media, and identity-making in relation to "organic" and explore the field of organic food as a whole. Thus, this book sheds light on the ways in which organic food is shaped and negotiated not only by organic farmers or organic food advocates but also by other social actors: entrepreneurs, consumers, journalists, dietitians, and legislators. It exposes the ways in which organic food shaped the subjectivities of these actors and also the ways in which the field of organic food was shaped by the habitus[86] these actors possess.

## Organic as a Glocal-Cultural Field

My analysis, which is grounded in the sociology of culture, draws on the works of those who study meaning-making and framing in a variety of modes of cultural production, especially those who examine how the aggregate actions of actors in fields of cultural-culinary and agricultural production generate and substantiate social categories.[87] It describes societal processes of translation and negotiation in relation to the meanings of the notion "organic" and how it is connected—symbolically and materialistically—to places and identities.

In order to understand the locus of these translational processes, I employ sociological notions from field theory. This approach allows us to understand "organic" as not only a set of agricultural procedures stemming from environmental ideologies but also as a product of "spheres of values."[88] In these social domains, the realizations of the notion "organic," as well as its "values," are constantly negotiated. I also draw on contemporary post-Bourdieusian accounts on fields of cultural production, which specify that social actors are always situated in more than one field and routinely transpose elements from one field of their actions and practices to another.[89] Thus, I designate how the meanings of organic food and its politics, aesthetics, and morality are all subject to tensions between actors whose work is structured by a cultural openness to global ideas and who are collectively striving to engage in global food-related trends and markets. Simultaneously, these actors use local interpretive frames[90] to position their own affiliation to local agricultural and culinary fields. In other words, I suggest seeing "organic" as organized around glocal fields of agricultural, culinary, and cultural production. These fields are contoured by growers, entrepreneurs, food manufacturers and grocers, global and local food distributers; local social movements and consumer associations; global and local creators of communication media, food literati, and cultural intermediaries intending to promote organic agriculture in the local sphere; global and local regulatory agencies, politicians, practitioners of "alternative" ways of living engaged with the global philosophical foundations of organic agriculture; and non-committed consumers who wish to use their "organic-reflective taste"[91] as a symbolic means for identity-making and social currency.

In the following chapters, I demonstrate how the philosophical foundations of organic agriculture, as well as facets of the global organic culture, are entangled with salient, local Israeli aspects: the ethos of *halutzim* ("pioneers"—Zionist ideological farmers and workers) (Chapter 1), the utopian visions of the Israeli kibbutz (Chapters 1 and 4), indigeneity that is claimed by both Palestinians and Jewish settlers in the Gaza Strip and the West Bank (Chapters 1 and 2), biblical meanings that have been ascribed to the (Western) "green" movement and counterculture ideas (Chapters 1, 2, and 3), and the Americanization of Israeli society and its neoliberalized economy (Chapters 3, 4, and 5). The chapters in this book are organized according to the analytic logic that directed the research that this book is based on, a logic that is divided into two dimensions: synchronic and diachronic.

Regarding the synchronic dimension, the chapters intend to demonstrate the multiple actors and agencies that work simultaneously on the translation and assimilation of the global notion "organic" in the Israeli-local context. Thus each chapter is focused on a different level and different actors operating in the Israeli field of organic food, for example farmers' movements and the "pioneers" of organic food in Israel as described in Chapters 1 and 2; food producers and consumers as described in Chapters 3 and 4; organic food distributers as presented in Chapter 4; and cultural intermediaries and regulatory agencies as discussed in Chapter 5. These chapters present a description of the field of organic food in Israel as a whole.

Regarding the diachronic dimension, the chapters are also arranged in a way that traces the development of the Israeli organic field over time. Thus, Chapters 1 and 2 focus on the birth of the field of organic food and its establishment in Israel through the 1980s and the 1990s; Chapters 3 and 4 describe the development and proliferation of organic food from the 2000s onward; and Chapter 5 indicates more recent developments, such as the discourse on organic food in popular media and the establishment of the organic law.

In Chapter 1, I argue that Zionist culture—idealized by the first farmers to start practicing organic farming in Israel—enabled the introduction of organic agriculture into Israel. I show how the Zionist (national) meanings attributed to agrarianism laid the foundations for the structural and symbolic restrictions of organic food within the cultural and political boundaries of Jewish-Israeli agriculture. This form of organic Zionism was expressed symbolically in the frequent use of biblical texts as narratives and justifications for the establishment of organic agriculture in Israel. From a structural perspective, this chapter shows that organic agriculture emerged as an agricultural niche designed as a means of strengthening the status of the Israeli agricultural sector in global agricultural markets and expanding its activities in them, and thus generally supporting Israeli conventional agriculture.

Chapter 2 explores the continuity of the nationalistic meanings attributed to organic food in Israel and demonstrates how global organic schemes can be implemented according to local ethno-national discriminative social logics. The chapter shows how the development of organic agriculture in the occupied territories (Judea, Samaria, the Jordan Valley, and the Gaza Strip) was rhetorically justified by agricultural-scientific discourse, arguing that these specific lands were cultivated by traditional agricultural methods (or not cultivated at all), uncontaminated and therefore ideal for organic farming. However, practically and ideologically, the Jewish organic sector in these areas conveyed neo-Zionist (neo-nationalist) meanings and new visions of "the redemption of the land," which draws boundaries between Jewish farmers and Palestinians farmers. Notwithstanding this, while organic agriculture has been loaded with ethno-religious traditionalism and neo-Zionist meanings along political lines, the notion of organic is being used as a signifier of post-Zionism and cosmopolitanism among the Jewish middle and upper classes in Israel.

While Chapters 1 and 2 discuss the sociohistorical development of organic in Israel/Palestine, as well as the economic and political levels of organic food, Chapters 3 and 4 focus on its contemporary cultural levels. Chapter 3 discusses processes of production and consumption of organic food as portrayed in the initiatives of community-supported agriculture (CSAs) and farmers markets in Israel. It is revealed that more than agricultural or agronomical hubs, CSAs and farmers markets serve as a means of demonstrating cultural openness to the world and are used to create an Israeli vernacular version of global "foodie" discourse and culture. These places, as well as the social dynamic inherent to them, contribute to the constitution of a new Israeli cosmopolitan and creative class. Through dealing with organic food, consumers and producers in Israeli CSAs and farmers markets demonstrate their creativity, technological skills, and digital literacy as well as neoliberal marketing, communication, and affective skills. Thus, young Israeli organic farmers and consumers use organic foodways to construct a distinctive hybrid identity that is comprised of global hip-agrarianism, new Israeli bourgeoisie, and the new Israeli ethos of a "start-up nation."[92]

While Chapters 1, 2, and 3 discuss the interactions between farmers, producers, and consumers, Chapter 4 focuses on organic food distributers as well as on their consumers. It starts by discussing the ways in which models of "corporate organic" traveled outside the Global North and were vernacularized in local contexts. It explores how grocery retail chains that branded themselves as "organic supermarkets" frame the notion of "organic" as they operate in the field of Israeli retail grocery chains. I show how organic supermarkets in Israel imported and assimilated the model of the American supermarket chain Whole Foods Market to not only "corporatize" organic agriculture at the structural level (i.e., transform its modes of production and distribution) but also "Americanize" organic at the symbolic level. The chapter continues by discussing the ways in which organic food is also distributed and marketed according to hybrid global-local cultural logic: the merger between global New Age culture and the agrarian-socialist ethos of the kibbutz (a communal settlement that held an important position in Israeli Jewish agriculture). Based on ethnographic research conducted in Kibbutz Harduf (in the Galilee, northern Israel), I show how organic food production in the kibbutz serves as a means of binding together modern-utopic communal ethos with a postmodern New Age lifestyle that promotes individualistic spirituality and an emphasis on narcissistic physical nourishment. I also explore the ways in which the symbolic aspects of the kibbutz's agrarianism are used as a means of commodifying the organic and how what started as a small-scale, sustainable, and local sector transformed into a corporate organic enterprise, promoting popular similes of the "kibbutz" and a "Galilee" taste. The case of organic food in Harduf reflects contemporary trends of privatization, individualization, and capitalization of the kibbutz in Israel as well as broader dynamics between market actors and countercultural movements and ideas.

Chapter 5 attends to the ways in which the meanings and practices of organic food are translated by actors who mediate between producers and consumers and operate in two external fields: the field of media and the political field of law. Drawing from in-depth interviews with cultural intermediaries and from thematic content analysis of media artifacts, the chapter explores how organic food is portrayed and valorized in Israeli mass media. It is argued that cultural intermediaries tie the notion "organic" to varied aspects of cosmopolitanism, to the political culture associated with post-Zionism, to a neoliberal understanding of sustainable consumption, and to self-care and other ideas that stem from global/ western health and natural foods movements.

The second section of Chapter 5 deals with the mediation process that has been taking place on the political plane, discussing in particular the Israeli Law for the Regulation of Organic Produce (which came into effect in 2008). Focusing on the sociohistorical processes that led to the enactment of the law, the chapter shows how states and institutional agencies translated global ideas related to organic agriculture and assimilated them into a local political and cultural context. I argue that the law is the outcome of a deflection of the economic and agricultural crisis in Israel (from the mid-1970s onwards) onto the emerging organic agriculture sector. This deflection eroded the boundaries between conventional agriculture and organic agriculture, thus paving the way for the formulation of a national organic law. Therefore, the Israeli organic law was first and foremost enacted and served as a means of developing an export-oriented Israeli organic sector and promoting the globalization of Israeli agriculture. Second, by establishing a legally recognized control over the notion "organic" and over the process of obtaining organic certification, the state demarcated Israeli organic agriculture within ethno-national boundaries and contributed to discriminatory practices in the access to organic and sustainable agriculture. The chapter concludes with a discussion about the ways in which cultural and institutional intermediaries delineate symbolic and pragmatic boundaries to the field of organic food in Israel. I argue that these boundaries demarcate the "organic" in Israel according to fundamental currents in present-day Israeli society: one toward a neoliberal political economy and cultural cosmopolitanism and the other toward ethno-nationalism.

The concluding chapter summarizes how the Israeli field of organic food, as explained throughout the book, provides a counterintuitive perspective on alternative agriculture and alternative foods and suggests that we need to understand the culture of "eating for change" and food activism as global discourses and practices. It emphasizes the important role that local culture and political economy take in the vernacularization and implementation of organic agriculture and other manifestations of food activism.

As this book elucidates, structural and cultural processes of globalization—which are usually attributed to industrialized food, the Americanization of culinary culture, or the proliferation of "ethnic" or "exotic" foods—are also embedded in foods that are associated with similes of alter-globalization as well as in the practices, tastes, and identities of those seeking to reform and transform the global food system. The following chapters illustrate how practices and representations of "counter-cuisine" and alternative farming are subjected to societal constellations of globalization and localization as well as pave the way to reinforcing broader conflicting processes of neo-nationalization and cosmopolitization. The glocal perspective that is suggested here permits us to understand both the structural and cultural paradoxes of organic food and agriculture.

## Notes on Methods

In his 1826 renowned book *The Physiology of Taste*, Jean Anthelme Brillat-Savarin claims that food-related issues are central elements of the social order. To him (according to sociologist Priscilla Parkhurst

Ferguson's interpretation),[93] the gastronomic field should concern every reader because it connects *individual taste* to *social context*. These exact three elements—"taste," "individual," and "social context"—directed the methodological and analytical approach of the research in this book.

The field of organic food is obviously different from the gastronomic fields that Brillat-Savarin talked about since "good taste" is not the only consideration that is at stake in this field. Previous scholarly works paid more attention to the production of organic, leaving the "taste of organic" somewhat overlooked. Indeed, much of the preceding literature separated out the levels of production, distribution, and consumption of organic food. Some scholars focused on the political-economic dimension of the emergence and changes in organic agriculture.[94] Others analyzed the discourse on organic food and the negations of this notion in social movements and political institutions.[95] And some treated organic food as a focal point of varied consumer choices, norms, and values.[96] This book attempts to present a comprehensive perspective, relating "organic" to interconnected material and cultural production. Using this analytical approach, I considered organic food an artifact subjected to competing agrarian ideas, norms, and values as well as concerns related to health, the environment, risk, cultural representations and identity-making, and of course, taste.

Brillat-Savarin's emphasis on the close connection between the "individual" and the "social context" is both inspiring and fundamental to any sociological analysis of food-related issues.[97] Therefore, I followed key individuals who were responsible for the realization of organic agriculture and of the meaning-making of "organic" in Israel: producers, consumers, members of organizations, actors in movements, directors of companies, industry members, spokespeople, media professionals, and institutional representatives. In thinking about organic as subjected to a *field*,[98] I was attentive to key individuals who translated and attempted to implement the philosophies that underlie the notion of organic as well as to representatives of organizations who constructed the field and currently operate in it.[99] Therefore, several sections in the following chapters focus on vocal and influential actors and seek to analyze the discursive content they produce and instill in the Israeli field of organic food. I focused on the distinction between the *explicit* symbolic expressions and meanings performed by key individuals in this field (i.e., the ways in which the notion of organic is expressed in practices and in ideologies and doctrines individuals attach to organic food) and the structural-institutional meanings that are produced *implicitly* through the operation of organizations, regulatory agencies, corporations, movements, media outlets, collective settlements, and the like, as well as through social practices (shopping, eating, communicating, etc.).[100]

With an attempt to better our understanding of present-day Israeli society—as much as it is reflected through the lens of Israeli culinary and agricultural fields—and to extend the sociological-theoretical debate on organic food, ethical eating, and alternative food movements by discussing the Israeli case, I utilized several methods. First, I conducted engaged ethnographic fieldwork (2008–2015) that included observations in organic farms, production plants, stores, markets (farmers markets and open-air markets), and restaurants. I also attended health food festivals, ecological festivals, and food and nutrition workshops. In addition, I conducted interviews, historical document analysis, and media content analysis.

Aiming to understand the emergence of the field of organic food in Israel, and the "social context" of its appearance, I analyzed a wide range of written sources, including dozens of publications from the IBOAA's archive as well as popular press articles published during the formative period of organic agriculture in Israel (1982–2000). I analyzed all published issues of the IBOAA newsletter produced from 1985 to 2000 under the titles *Renewable Agriculture* (*Haklaut Mithadeshet*), *Organic Agriculture* (*Haklaut Organit*), and *Organic* (*Organi*). The issues of these journals—which were distributed to members of the IBOAA and, in a limited way, to consumers—revealed the discourse that took place between organic farmers and state institutions, and later, following the expansion of the food field in Israel, the discourse that took place between farmers and consumers. Most of my findings based on this method are included in Chapter 1.

In order to understand the trajectory of the field's changes and dynamics, I also analyzed contemporary written sources (published between 2000 and 2019). Among them are industry and trade publications, market research studies, governmental policy papers, and research papers. I collected a variety of cultural material related to organic food, including print, digital, and visual items. In addition, I analyzed websites of farms, retail chains, CSA initiatives, farmers markets, kibbutzim, governmental authorities, and private certification companies. I also conducted thematic content analysis of newspaper columns dealing with food and sustainability in Israel ($N \approx 300$ columns) taken from three daily newspapers, five popular environmental and lifestyle magazines, cookbooks with direct references to organic food, nutritional handbooks, and tourist guidebooks. This latter analysis helped me understand how the public image of organic food in

Israel has changed over time. In addition, I analyzed 34 protocols from the Knesset Economic Affairs Committee meetings—where the discussions of the Law for Regulation of Organic Produce took place and the main legislative process was carried out. I consulted numerous drafts of the bill, the memorandum of the law, and the law itself. Findings from this analytical method are discussed in Chapters 3, 4, and 5.

The interviews I conducted were intended to fill in aspects of the history of organic food in Israel that are not in the written record, as well as to better understand the experiences and motivations of those working in organic food production and those who consume organic food in Israel. I conducted 35 in-depth, semi-structured interviews with organic food producers in Israel: farmers, growers, factory owners, cooks and restaurant owners, retail chain managers, store owners, and representatives of organic food packaging and distribution companies. I also conducted 11 in-depth interviews with cultural intermediaries operating in this field: journalists, media professionals, and nutrition consultants working to promote organic food production and consumption in Israel. I interviewed three officials from the Ministry of Agriculture and Rural Development (MARD) who were involved in the legislative process and the formation of the Law for Regulation of Organic Produce. In addition, 45 interviews were conducted with organic food consumers. Those interviews included in-depth, semi-structured interviews, most of which took place between 2009 and 2012, with people who testified to consuming organic food on a regular basis. I also conducted informal interviews with consumers whom I met at the various sites where organic produce is sold. Most of my interlocutors were of the middle and upper middle classes, many of whom worked in the fields of law, engineering, education, high-tech, media, advertising, art, or other creative industries, or held other professional-managerial roles. Others were employed as institutional or private therapists in the conventional health sector or in the alternative medicine sector. Findings and insights from all of these interviews and discussions, which were thematically analyzed, appear throughout all of the chapters. Overall, these diverse methods, as well as the analytical approach that looks on both the structural-institutional level and the symbolic-expressive level, helped illustrate the societal dynamic in the field of organic food and the process of meaning-making for organic food in Israel.

Chapter 1

# "If Organic Agriculture Is Here—
# Homeland Is Here"

## The Birth of Organic Agriculture in Israel/Palestine

> If workers of the land are here——a nation is here, a homeland
> is here.
>
> —Smilansky, 1926: 3

This chapter explores the social conditions within which organic agriculture in Israel emerged, and the ways in which organic agriculture in
Israel was articulated in its formative years.

Although the story about the ways in which the field of organic
food in Israel has evolved—as will be presented in the following chapters
and throughout this book—suggests that this field has been shaped by
changing national ideologies, the origins of organic agriculture in Israel
are rooted in the long history of the Zionist settlement movement and
its relationship to agriculture.

Agriculture in Israel has always been tied to national ideologies,
and for decades it has been politically supported by the state.[1] Take for
example the renowned aphorism by Moshe Smilansky (1874–1953, a
farmer and a national leader) quoted in the epigraph above. This saying
has been, and is still, learned and quoted by almost every Jewish farmer
in the Land of Israel for nearly 100 years. Although shortened from the
original, the saying "If agriculture is here—homeland is here" is ubiquitous. It is written prominently on the wall of the building housing the

Farmers' Federation of Israel (Hitahdut Haikarim) in Tel Aviv and on their website. It is displayed on grain warehouses and packing houses in many farms in Israel. It is repeated by every politician who promises any sort of support to the Israeli agricultural sector. In recent years, demonstrations and protests by farmers demanded a restoration of the link between the Israeli agricultural sector and national resilience, drawing on Smilansky's words to push forward an idea that has been prevalent in Israel for many decades.

Indeed, agriculture has been a central ethos since the early days of the Zionist (Jewish nationalism) project in the late nineteenth century and the emergence of Israel as a modern state. Farming, as a practice and an ideology, played an important role in the Zionist-Jewish immigrants' reclamation of a Jewish national birthright as an indigenous people returning to their promised land. It was perceived as a symbolic connection between the nation's past and its future, as a tool for land-holding, and as a means of political and economic independence.[2] The notion of "working the land" was interwoven with various cultural aspects throughout the history of modern Israeli-Jewish society, especially among those who emphasized the connection between the Jewish people and the Land of Israel in the national-historical narrative. An early expression of this can be seen in the way agricultural representations and practices were used to cultivate the image of a group known as "idealistic pioneers"—the halutzim. In Israeli historiography, the term *halutzim* refers to a group of ideologically driven young men (and a small number of women) who arrived in Palestine primarily from Europe between 1903 and 1914 and aspired to engage in agriculture. They perceived themselves as responsible for the "conquest of labor," namely the entrenchment of the Jewish worker in agricultural work on Jewish land in the service of the national Jewish project. "Working the land" was believed to bestow upon the *halutz* (an individual pioneer) spiritual qualities and a mystical connection with the country and the land. It was through agricultural labor that the halutz was redeemed from the heritage of exile and the entire group of halutzim redeemed the nation.[3] Thus, *halutziyut* (pioneering) and *hityashvut haklait* (agricultural settlement) jointly played a constitutive role in the Zionist settlement project and in shaping Zionist culture,[4] as indicated, for example, by the image of the sabra (*tzabar*)—a Jew who is born in the Land of Israel—which was shaped by extensive use of agricultural images.

On the political-economic front, this affinity between agriculture and "redeeming the nation" led to the establishment of collective agricultural settlements of halutzim on lands purchased through national funding (beginning in 1909), which was a strategy meant to guarantee a Jewish hold on the land and monopoly over labor. After World War I, these agricultural settlements developed into the forms of kibbutzim (communal settlements that were traditionally based on agriculture) and moshavim (cooperative smallholders' villages), which were affiliated with *halutzi* movements. Those Jewish agricultural settlements enjoyed a privileged status and access to national capital and laid the foundation for developing a modern-Western agricultural sector. Later, the State of Israel (established in 1948) set out to expand Jewish agricultural production and provided widespread cultural and economic support.[5] Agricultural production in Israel grew exponentially over the years, except in the case of Palestinian farmers who found themselves dealing with severe limitations on access to agricultural resources. During this time, the state appeared to be more conducive to larger, rather than smaller, agricultural operations and was inclined toward export markets.[6]

Since 1967, agricultural practices have been integrated into the Jewish settlements in the Occupied Territories, and the overall agricultural production has increased. Due to the economic well-being that prevailed after the War of 1967 ("The Six-Day War") among several groups in the Jewish population in Israel, the relative expenditure on food (and especially on agricultural produce) for private consumption decreased.[7] As a result, the sector's profitability from local sales decreased, and the sector began to rely on foreign markets. In the 1980s, the Israeli agricultural sector faced a severe crisis due to over-investment and the inability of many farmers to repay their debts, a result of cheap credit given by the state.[8]

Between the mid-1980s and the end of the 1990s, Israeli society went through a profound socioeconomic change, including economic (neo)liberalization and adaptation to economic globalization, technological change, and privatization.[9] These political and economic changes affected the agricultural sector. Values such as solidarity and collective commitments—which were associated with Jewish-Israeli agrarianism—gave way to the logic of the global market, and thus the agricultural sector was denied the preferential economic treatment it had enjoyed during the country's first 30 years. Profit, rather than ideology, lay at the heart of Israeli agribusiness decisions. Integration into foreign agricultural markets was conceived as the main means of increasing economic profits, and thus high-tech agricultural branches with high demand in the export markets began to flourish.

Corresponding with these trends, agricultural research and development boomed, and agricultural niches for export were encouraged by the state, which saw this as a source for improving reserves of foreign currency and as a solution to the crisis that has faced Israeli agriculture since the 1980s.[10] Since then, Israeli agriculture has been moving in the direction of mechanized and industrialized agriculture, which is based on high inputs and heavy usage of chemical pesticides and inorganic fertilizers.

Organic agriculture—at least in its ideal form and according to the philosophical foundations described in the introduction—stands in particularly sharp contrast to these processes. It could be expected that the flourishing of Israeli hyper-industrialized conventional agriculture—which has turned out to be disconnected from Israeli sociocultural life and abundant in health and environmental hazards—would serve as a "fertile ground" for the rise of counterreaction in the form of alternative and oppositional agriculture, such as organic agriculture. As this chapter suggests, the emergence of organic agriculture in Israel occurred alongside, and not divergent from, the industrialization and globalization of Israeli agriculture. As I will describe below, organic agriculture in Israel has come to play a role in the challenges the Israeli agricultural sector faced, including coping with global-economic trends and cultural changes in the ways in which Zionism is reflected in agricultural practices.

Several pathways converged to allow Israeli organic agriculture to rise in conjunction with conventional agriculture. But no one played a more central role in these developments than Mario Levi, also known as the pioneer (*halutz*) of organic agriculture in Israel. Levi's personal story, as described below, is an emblematic illustration of the ways in which the field of organic food emerged and evolved.

## The Halutz of Organic Agriculture

I met Mario Levi on a sunny Sunday morning in March 2011 in Sde Eliyahu, a religious kibbutz in the Beit She'an Valley (in the northern part of Israel), where he lived. During the phone conversation I had with him on the eve of our meeting, he gave me the following driving directions: "Drive 500 meters until you see a sign that says, 'organic vineyard.' Continue straight until you see a big tree and under it a shack. I'll meet you there!" I followed his instructions. Under a thick tree, I noticed a rickety shack. The door of the shack was decorated with a large Star of David (the symbol of modern Jewish identity and

Figure 1.1. Levi's shack, Sde Eliyahu. Photo by the author, March 2011.

Judaism) and a big sticker showing the letters IBOAA—the acronym for the Israel Bio-Organic Agricultural Association (see Figure 1.1). When I arrived, I saw Levi, then 87 years old, dressed in his work clothes with a hat on his head. He was bending over a raised bed near the shack, gardening. The old shack, the aged tree, and Levi's smiling and wrinkled face blended in my imagination into a visual composition that reflected longevity, rootedness, and earthiness. The entire scene seemed to me an iconic spectacle of organic agriculture.

Notions of rootedness and earthiness also came up in his remarks during our conversation. Before I even managed to introduce myself and address the questions I had prepared for the interview, he began the conversation by lecturing me about what organic agriculture meant to him, using these notions:

First of all, we should understand what we're talking about. What is organic agriculture? It is an agriculture that sees nature. It is rooted in humanity and in the environment. It is *connected to the soil* and produces the best from it in a proper and healthy way. It is an early agriculture that for many years provided a healthy and long life for human beings. In my opinion, it will extricate humanity from the pit in which it finds itself. From an ecological perspective, and according to the Torah's commandment to search for the truth, I found that organic agriculture has pioneering power [koach halutzi] for the benefit of all human beings, and especially for the people of Israel.

Indeed, pioneering, power, and vitality arise from Levi's story and from his persona, corresponding perfectly with the myth and ethos of Israeli halutzim. Levi, who passed away in 2018 at the age of 94, was born in Trieste, Italy. He immigrated to Palestine in 1939 as part of the Youth Aliyah (Aliyat Hano'ar)—a Jewish Zionist movement that rescued young adults from the Nazi regime and arranged for their resettlement in Palestine in kibbutzim. The organization also operated in Italy, and Levi was one of the 40 young people who managed to settle in Palestine. Upon his arrival, Levi was sent to Mikveh Israel, the first Jewish agricultural school in this region. After being trained as a modern farmer and learning Hebrew, he was sent to Kibbutz Sde Eliyahu (established in 1937 on the lands of Arab al-Areeda, then a Palestinian Arab village). In an unpublished memoir that he wrote, Levi described his feelings in those times using a typical Zionist-pioneer tone: "We understood that the advancement of the Jewish people in the Land of Israel was rooted, first of all, in settling the whole country, and especially the most remote places. All I wanted was to join one of the most important tasks of the Jewish people in our generation."[11]

During the 1960s, the second decade after the establishment of the State of Israel, Levi was appointed director of vegetable farming in Sde Eliyahu—at that time the kibbutz's main agricultural branch. Among other things, he was involved in growing bell peppers ("California Wonder Bell Pepper") and was responsible for selling the produce for export to Europe. The bell peppers grown for export flourished until pests caused considerable damage to the crops. Levi and his colleagues frequently used pesticides, but to no avail, and their pepper agribusiness was dropped. His sense of failure in this endeavor led him to doubt the efficacy of pesticides and synthetic fertilizers. He expressed his doubts to various people at the Ministry of Agriculture (MARD)[12] with whom he worked, while searching for other branches of agriculture.

At the same time, biologist Hans Müller—one of the founders of bio-organic farming in Switzerland who influenced the spread of organic agriculture to many countries[13]—visited MARD after hearing about Israel's agricultural development and achievements. He wanted to introduce bio-organic farming—which was unknown in Israel at the time. Most of MARD's staff seemed reluctant to discuss the matter, but a few of them, who knew Levi's skeptical view regarding conventional agricultural methods, suggested that the two of them should meet. And so it was.

In 1969, Levi was sent by MARD to participate in a bio-organic agriculture course in Switzerland led by Müller. He was sent there as part of MARD's search for new agricultural niches. In the report he submitted to MARD upon his return to Israel, Levi stated that he believed in the feasibility of organic agriculture in Israel. He further elaborated that there was a great demand for organic food in the European markets, and that this field may contribute to the development of agricultural export from Israel. However, people from MARD were not convinced by his reasoning and refused to cooperate or even support the idea. Nevertheless, Levi was determined (or he had no other choice) and decided to try to promote organic agriculture on his own.

" 'Nothing stands in the way of desire'—this statement has been proven repeatedly in our kibbutz," Levi wrote in his memoir,[14] describing those early days when he tried to establish organic agriculture, connoting here the key elements of the halutz ethos: realization (*hagshama*, namely the implementation of one's Zionist and personal ideals) and a desire to fulfill a pioneering "life project" and "personal vision," while engaging in agricultural labor.[15] However, what helped Levi in the realization of his vision was not only his willpower but also—and perhaps even more importantly—the institutional framework and the resources of the kibbutz in which he operated. The kibbutz had the ability to provide lands and other agricultural resources, as well as economic protection against market forces and social solidarity—all of which are essential, of course, to the implementation of pioneering ideas, such as those Levi had. Thus, he asked his friends, the kibbutz members, to support him in his efforts to develop organic agriculture as a new agricultural branch instead of continuing with the "bell pepper for export" branch that had failed. In an interview, he told me,

I thought it could be the solution for some of the problems the kibbutz encountered. I had to convince the kibbutz members to do something that nobody in Israel did before. I told them: It works elsewhere in the world, why shouldn't we be "a nation like any other." We could even be leaders in this agricultural sector and be a "light unto the nations."[16]

The kibbutz members, contrary to Levi's colleagues from MARD, were convinced and provided him with the resources he needed. Today, about five decades later, one of the great prides of the kibbutz members is their role as "the pioneer of commercial organic farming and biological pest control in Israel"—as noted on the kibbutz visitor's center website.[17] The kibbutz website also claims, "Many of our field crops and fruits are special in that they are cultivated according to the principles of organic agriculture. We were real pioneers in this field in Israel fighting for the exclusion of toxic fertilizers and sprays."[18]

Back in the 1970s, when Levi's first attempts at small scale organic farming succeeded, he invited senior officials from MARD to come and see. At that time, they were impressed, but still, their willingness to support Levi in developing organic agriculture at the national level remained minimal. Nevertheless, they did encourage him to continue with his organic methods for two reasons. First, they considered the plot of land cultivated by Levi using organic methods to be a sort of "agricultural laboratory" or "control unit" for (conventional) agricultural research. Second, since Israel was perceived as a "world power" in agricultural research and development, MARD received many requests for international cooperation, among them requests and queries related to issues of biological pest control methods (an issue that aroused great interest, especially in the Global North, where concerns about the implications of the use of synthetic and chemical pesticides and fertilizers began to take shape during the post–*Silent Spring* era).[19] Thus, the interests of Levi and his friends from Kibbutz Sde Eliyahu (who needed alternatives to their afflicted export branches) were in line with the interests of MARD—which strived for the promotion of research and development for the benefit of Israel's conventional agro-industry sector.

Only a little while later, when Levi proved that he could produce organic crops on a larger scale, did MARD begin to discuss with him the possibility of institutionalizing organic farming into an agricultural sector under their auspices and support. In fact, the development of the Israeli organic sector, which would offer organic goods to Israeli consumers, was perceived as an investment with low potential for economic profits, and thus impractical. Instead, MARD encouraged him to think

Figure 1.2. Organic pomegranate vineyard, Sde Eliyahu. Photo by the author, March 2011.

about the possibility of further developing an export-oriented organic sector. And so, he did.

## "We Developed an Organic Export System that Should Be a Source of National Pride"

In 1982, Levi took the first step toward institutionalizing an Israeli organic sector by establishing the Israel Bio-Organic Agriculture Association (IBOAA). This association included a number of farmers who had decided to take part in growing organic produce for export. The formulation and regulation of organic agricultural methods was done by members of the organization. In order to ensure that what the Israeli organic farmers produced was indeed organically-grown, Levi and his colleagues from the IBOAA adopted a system of peer review and self-regulation—a system that was common elsewhere in the world.[20]

As a second step, Levi got IBOAA accepted as part of the Europe-based International Federation of Organic Agriculture Movements

(IFOAM), which began to develop international standards and guidelines for organic agriculture around the world. It turned out to be a meaningful step, since in 1978 IFOAM created standards that later served as the basis for organic certifications and national legal standards, such as EC Reg. 2092/91 (EU legally enforceable and officially recognized standards for organic crop production, certification, and labeling) as well as the American federal organic law and the National Organic Program (NOP) in the United States.[21] Levi and his colleagues from IBOAA became involved in working with their counterparts from IFOAM, and some of them were engaged in the formulation of EU and IFOAM organic standards. In this way, Israeli organic farmers played an active role in both creating the globalization of organic food and positioning Israeli organic agriculture in what was then a developing global market. Thus, from its inception, Levi designed organic agriculture in Israel for export and global markets.

Third, Levi created a close and extensive collaboration with the government-owned company Agrexco (founded in 1956). This company played a central role in connecting Israeli agriculture to global markets. Originally, the company was responsible for exporting surplus, fresh, and local agricultural produce. For four decades Agrexco enjoyed a monopoly on the Israeli agricultural export market, since all exports of agricultural produce had to be done through the company.[22] Using his Zionist habitus (to use Bourdieu's terminology [1984]), his social and cultural capital as a Zionist ideologist farmer, and taking advantage of his long acquaintance with the Israeli agricultural organizations, Levi managed to convince Agrexco—the flagship of conventional agriculture in Israel—that it would be worthwhile for them to export organic produce to Europe. In the interview I conducted with him, he said,

> At first Agrexco did not want to deal with organic at all. But I *knew* how to talk to them. *I was one of them* [his emphasis]. I had friends there, from my time in Mikveh Yisrael. We were not just talking business. I spoke with them about Zionism, I asked them not to "look only through the hole of the Qirsh" [he used an expression in Hebrew that literally means "not to think only in the monetary aspect"]. Eventually, they understood that this is the right thing to do, and that it could also bring economic benefit. I convinced them that it would be possible to export organic produce to Europe at a high price and that we would be able to create a demand for organic

produce grown in Israel. After all, Israeli agriculture has an
excellent reputation. This is how we developed an organic
export system that does not disgrace Israel and actually should
be a source of national pride.

Since then—more than two decades—almost all organic produce grown
in Israel has been exported. In 2009, Agrexco reported a sharp increase
in the export of organic produce to Europe (especially to Germany, the
UK, France, Italy, and Switzerland). Sales amounted to 550 million
euros.[23] The regulation of organic produce intended for export was
performed by the Plant Protection and Inspection Services (PPIS) of
MARD as part of their broader responsibility to enforce regulations of
all agricultural products exported from Israel. This collaboration between
those three organizations—the IBOAA (which was still in its infancy),
Agrexco (a conventional agriculture, government-owned company),
and PPIS (a governmental regulation authority)—succeeded beyond all
expectations.

Subsequently, a whole unit was established in Agrexco to specialize in exporting organic products, operating under the brand "Carmel
Bio-Top." During the early 1990s, before the process of the "conventionalization of organic" in which organic movements in the Global
North were co-opted by conventional food sectors,[24] this collaboration
between organic farmers and conventional agricultural organizations
was conceived as quite unusual. For example, a representative of the
IBOAA, who participated in an IFOAM conference, reported in the
IBOAA newsletter: "This is an exceptionally remarkable phenomenon
where a fairly well-known conventional food company that specializes
in marketing conventional agricultural produce at the same time exports
produce with organic standards."[25]

Thus, while in 1993 the export of organic produce from Israel was
estimated at 6,400 tons and the volume of trade was estimated at ₪40
million (US$11 million),[26] in 2008 the export was estimated at about
80,000 tons and the sales turnover was estimated at ₪1 billion (US$270
million). Agrexco exported almost all of the Israeli organic agricultural
produce during those years.[27] Levi explained this increase in organic
export using typical halutzi terminology: a group of visionaries facing
dire straits, eager to overcome the obstacles and to "conquer the land,"
succeeded against all odds in overcoming difficulties and creating pragmatic revolutions that significantly contribute to the collective and the
nation. In the interview he told me,

The principle was to export as much as possible. It was important to us, first of all, to expand the market, to increase the number of organic farmers in Israel. I persuaded our best farmers to join. These were people who had vision and passion. They not only saw that it would be profitable for them personally, but they also understood the importance of organic, they realized its great contribution to the nation.

One-by-one, farms engaged with organic-for-export agriculture were established throughout the country, including a large-scale organic farming system in the Jewish settlements of the Occupied Territories in Judea, Samaria, and the Gaza Strip (see Chapter 2, which deals with neo-Zionism and organic agriculture in the Occupied Territories). Levi was filled with a sense of patriotism as he talked about it:

This is one of my greatest sources of pride! A few years ago, I went to the Hebron Hills. We reached Efrat [a Jewish settlement]. I went into a huge organic store near an organic field. I saw dozens of consumers there. There was hardly anything here 20 years ago. I literally cried with excitement. I couldn't help myself. Our efforts and strenuous work were not for nothing! I thought to myself, it's just like the famous saying [of Smilansky]: "If agriculture is here, homeland is here." Israel's organic agriculture thrives, and it takes part in this whole process of strengthening the connection to the land, to the Jewish settlement project, to Zionism and hard work! It moved me in an extraordinary way.

Levi had worked tirelessly to gain legitimacy, recognition, and support from Israeli agricultural institutions, and his work paid off in the form of a thriving export-oriented network.

Although he boasted about the "national pride" embedded in the Israeli organic sector that developed, almost all the organic produce from Israel did not reach the local Israeli market (which is at odds with one of the important aspects of the organic idea: locality). The export orientation of organic agriculture in its formative years is evident, for example, in Levi's report from 1993 addressed to the IBOAA members in which he summarized the agricultural activities and achievements of the previous year:

A baby food company in Germany ordered three times more frozen corn for next year; a huge food company operating in Italy is about to order thousands of tons of processed and frozen vegetables and fruits. We are negotiating with representatives of a large German agricultural cooperative, which works with a big German retail chain, where there is demand throughout the year! I could go on with many other examples. . . . Today Israel is a powerful exporter of organic agriculture.[28]

The extensive preoccupation with international trade is also evident in the plentiful articles published in the IBOAA newsletter dealing with organic around the globe. For example, articles that described organic consumers' profiles in Europe, Australia, Japan, and the US; a monthly column bearing the title "Organic Around the World"; reports and reviews of the activities of Levi and other IBOAA representatives in international conferences and their efforts in marketing Israeli organic produce in global markets; and much more.

The legitimacy and support given to the activities of IBOAA by Israeli (conventional) agricultural institutions was due to the growing volume of exports. Organic agriculture functioned as an agricultural niche, one among other agricultural niches that was encouraged because of its potential to strengthen Israeli agriculture as a whole. For example, at a 2006 conference dealing with organic agriculture, Professor Dan Levanon, who served as the Chief Scientist of MARD, announced, "Organic agriculture, which is export oriented, is definitely a growing sector, and this is very important, because we think that the growth potential of Israeli agriculture lies mainly in exports."[29] Hence, the driving forces behind the evolution of Israeli organic agriculture were the attempts to integrate it in the global organic trade, while strengthening conventional agriculture. When I interviewed Levi and provocatively asked him if the extensive preoccupation with organic exports contradicted aspects of organic ideals (for example, responsibility for reducing environmental pollution and carbon footprint resulting from accelerated activity of transporting food products throughout the world), he replied firmly,

It is a false and misleading assumption that an organic farmer should only operate on a small scale. Why?! If all of humanity is in trouble, and there is a group of farmers in Zion that can help Europeans or Americans who realized they should eat non-conventional fruits and vegetables, is it so bad to sell them Israeli organic produce? Not at all! I must tell you, I felt pride every time I saw a box of organic vegetables in any of the European supermarkets, as the saying goes: "for out of Zion shall go forth the law" [referring to the biblical eschatology from the Book of Isaiah].

This biblical quotation was not the only one I heard from him. In fact, he—and many of his colleagues, "the Israeli organic pioneers"—frequently used biblical texts and metaphors as a means of justifying the establishment of organic agriculture in Israel. Thus, while on the political-economic structural level they were occupied with export and dealing with "the globe," on the symbolic level, as described below, their actions bore local, Jewish, and Israeli meanings.

Figure 1.3. Bio-Tour: Levi at work—wall painting, Sde Eliyahu. Photo by the author, March 2011.

## Symbolic Establishment—
## Jewish-Israelization of Organic Agriculture

The sociological usage of "field" as an analytical tool refers to a set of social activities (production, distribution, marketing, consumption, etc.) that become circumscribed by a specific symbolic and discursive framework.[30] The progenitors of organic agriculture in Israel realized that dealing with economic and institutional tasks would not be enough, and that they had to make an effort in establishing a symbolic definition and cultural conceptualization of "Israeli organic." For many years, Levi and his colleagues endeavored to pour symbolic meanings into organic agricultural practices. Biblical verses and images from Zionist history and ethos served as the main components in Levi's "cultural tool kit."[31] Whenever he tried to persuade conventional farmers to make the transition to organic, Levi did not discuss just the procedural or instrumental aspects of the matter and did not only use universal ideological claims (global environmentalism, for instance). While he did make use of these concepts, he mainly emphasized aspects of national-Israeliness and spiritual-religious-Jewishness. "One cannot look at organic agriculture only in a narrow perspective," he would preach to his listeners. "Organic agriculture is not 'just' one type of agriculture out of many; it embeds a human approach and especially a Zionist and Jewish approach." This is what Levi expounded in the speeches he delivered at various agricultural conferences and at the many national agrarian symposiums he attended.[32]

During the establishment of IBOAA, Levi gathered around him a group of farmers who worked hard to formulate a "gastronational"[33] (or rather "agronational") conceptualization of the notion "organic." Yossi, who served for many years as the main instructor in the IBOAA's course "Basic Organic Agriculture" (the main platform for training new organic farmers in Israel and for converting industrial farmers to organic), used to open his lectures with the following statement: "Organic agriculture is actually a Zionist agriculture! Guess what, one of the founders of organic agriculture is no less than Lady Eve Balfour, who was related to Sir Arthur Balfour, yes, the renowned one from the Balfour Declaration."[34] Yossi said this jestingly, but nevertheless it created an associative and symbolic coupling between a central event in the history of modern Israel (the Balfour Declaration) and organic agriculture. Humor, according to sociological accounts, cements social order and knits groups in a shared understanding, making a common cause evident through collective amusement.[35] Therefore, if we seriously consider the ramifications of this statement on the indoctrination of new Israeli organic farmers, then we should conjecture that these humoristic sayings instilled revolutionary, pioneering, and national-Zionist meanings in the emergence of Israeli organic agriculture.

Another, similar example can be discerned in an article in the IBOAA's newsletter describing the establishment of an organic farm in the northern part of Israel:

> Yoel Moshe Salomon[36] preached the commandment of set-
> tling the Land of Israel. He was one of the founders of the
> agricultural settlement in the Land of Israel. Guy, the son of
> Yoel Moshe's great-grandson, continues this commandment
> [mitzvah] and nowadays cultivates land according to the prin-
> ciples of organic agriculture. His grandfather dealt with the
> redemption of the lands of Israel and purchased the Makura
> Farm from the Arabs. . . . Guy cultivates mainly olive groves.
> After realizing that the oil mills in the area, operated by the
> Arabs living in the area, were unable to produce high-quality
> olive oil—he purchased machines from Italy and established a
> cutting-edge and sophisticated organic oil mill, which became
> a tourist attraction. . . . The olive oil from Makura Farm is
> sold in bio-stores [prestigious natural food shops] in Tel Aviv
> in designed bottles. And what about the quality of his olive
> oil? Well, it is the best in the country![37]

Here, too, organic olive oil production is described by the authors of the article using symbolic representations of the Zionist historical narrative: the mythical story of Yoel Moshe Salomon, which played a part in the Zionist historiography; notions such as "redemption of the land"; and stories about the alleged superiority of Jewish agriculture over the local Arab agriculture. All these representations were used to make symbolic connections between organic agriculture and Zionism.

Within the Zionist-halutzi ethos, the Bible is the seminal text. It nurtured the sense of homeland and symbolized the connection with the national past, a source of national pride, proof of Jewish connection to the soil of the homeland and as a means to espouse the love for the Land of Israel.[38] Therefore, Levi and his colleagues used a myriad of biblical and Jewish-religious sayings and representations when they discussed the theory and philosophy of organic agriculture and when they instructed farmers on organic practices. For example, the curriculum of IBOAA's

"Basic Organic Agriculture" course included a lecture by Levi entitled "The Inextricable Connection between the Ecological-Organic Concept and Rooted Judaism." This lecture began with the following:

> The understanding of what was supposed to be considered as "nature" has been disrupted. But if most of human beings have forgotten, the Jewish people must not forget. As it is written [quoting from Midrash Kohelet Rabbah]: "When God created the first human beings, God led them around all the trees of the Garden of Eden and said: 'Look at my works! See how beautiful they are—how excellent! For your sake I created them all. See to it that you do not spoil and destroy My world; for if you do, there will be no one else to repair it.'" We have faith, we must understand, we have to take care of what was created. "The LORD God took man and put him in the Garden of Eden to work it and take care of it" [quoting from Genesis 2:15]. . . . I explicitly call upon the rabbinate to join this holy mission, which in my humble opinion is no less important than any other commandment.[39]

All of the biblical and Israeli-Jewish historical images that dominated the formative period of organic agriculture in Israel served as a way of binding it to the historical Zionist narrative that was characterized by a yearning for the Land of Israel and the hybridization of nationalism and religion.[40] Yet, while these *local-national* cultural representations worked to attract farmers and establish the symbolic legitimization of organic agriculture in the Israeli agricultural field, from a structural perspective, as we discerned, the sector of organic agriculture was established as a *global-post-national* sector with a strong bias toward exports and international markets.

The establishment of organic agriculture and the early stages of the formation of an Israeli field of organic reflect on the multiple levels in which global ideas and practices—such as the global ideas related to organic farming—are crystalized into local fields of material and cultural production. On the cultural-symbolic level, the notion "organic" was entwined with Israeli national ideology and myths, particularly the ethos of Zionist halutzim. This intersection between nationalism and agrarianism was used to justify the establishment of what is supposed to be a field of alternative food production. Yet fields of food production that are imbued with "alternative" meanings were not only developed by the countercultural undertaking of "back to the land"; they also grew out as agricultural niches designed to support the state and conventional food sectors in their efforts to gain a foothold in international markets and for dealing with economic and political changes.

Over the years, organic food in Israel ran along parallel global and local tracks. In the next two chapters I will describe both the continuity of the nationalistic meanings attributed to organic farming and its ascription to global-cosmopolitan meanings. The following chapter discusses how organic ideas and practices have served as a form of ethno-nationalism and ethno-religiosity. Further, it explores the ways in which organic agriculture has been entangled with one of the most salient aspects of modern Israel: the settlement movement and the acts of Jewish settlers in the Gaza Strip and the West Bank.

Chapter 2

# Organic Food Across Borders

## The Politics of Agriculture and
## Jewish Fundamentalism in the Occupied Territories

But the land, whither ye go over to possess it, is a land of hills and
valleys, and drinketh water as the rain of heaven cometh down.

—Deuteronomy 11:11

We don't have any connection with our neighbors [the Palestinians
farmers].

—Erez, an organic farmer from Rechelim,
a religious community in the hills of Samaria[1]

From the mid-1980s until 2005, about 50 Jewish farmers engaged in
organic agriculture, operated in the Gaza Strip area. Together with dozens
of other conventional farmers, they coalesced as a network of Jewish
farms beyond the Green Line. In the summer of 1997, the headlines of
Israeli mainstream newspapers screamed "Organic Fraud in Gush Katif"
and "The Gush Katif Scandal." It was reported that conventionally grown
produce that was claimed to be organic was exported to Europe from
Gush Katif, a bloc of 17 Jewish settlements that were located in the
Gaza Strip. "This is one of the biggest agricultural scams in the history
of Israel: farmers sent hundreds of tons of regular [conventional] vegetables to England, Germany and other European countries and labeled

45

them as organic crops," reported the daily financial newspaper *Globes*. It was also asserted that "produce that was disqualified in Europe due to the detection of pesticide residues was sold by these farmers in the local organic market."[2] The charges were serious: Claims were made that these farmers used pesticides prohibited for use in organic farming. "They allowed themselves to work with dangerous and toxic substances such as methyl bromide," explained one of the farmers in Gush Katif to the reporter who covered the story. The farmer continued with, "[Methyl bromide] is strictly prohibited in organic farming. [These farmers] lied brazenly."[3] According to the publications, these farmers—Jewish settlers from the Gush Katif settlements—secretly purchased large amounts of non-organic vegetables grown in the Mawasi (fertile lands in the southern part of the Gaza Strip cultivated by Palestinians). "The vegetables," it was reported, "underwent '*organic Jewish conversion*': they were put in Israeli boxes and exported as Israeli organic produce, sold in the European markets at double the price and triple its real value. . . . [These farmers] laundered Arab produce."

As seen in the *Globes* article, the discourse that evolved around the "Gush Katif scandal" exposed some of the common terminology among many farmers in Gush Katif and in other places in the West Bank. This terminology seems to be reminiscent of Mary Douglas's analysis of purity and danger:[4] the association of organic agriculture with political symbolic meanings of "Jewish purity" and as essentially separate from the "impure" local Arab agriculture. The investigative coverage and the exposure that some Jewish farmers "polluted" the Jewish organic produce was detrimental for some time to the organic agriculture enterprise that developed in Gush Katif. Soon after, however, the organic food businesses in that area recovered, and they continued to flourish until the Israeli disengagement from Gaza in 2005.[5] After all, organic agriculture was one of the region's main sources of employment. However, the quick recovery from the crisis was probably due to the fact that the Jewish organic agriculture enterprise that operated in Gush Katif embodied not only economic significances and meanings but also symbolic, religious, and national ones. In 2005, Yigal Kamintzki, who served as the chief rabbi of Gush Katif, said on the eve of the disengagement, "The settlement [movement] in Gaza Strip [i.e., in Gush Katif] is a reflection of an entire Jewish life, of Torah and labor. . . . [It is] an area that despite all the difficulties is a leader in the export of cherry tomatoes, herbs, and organic vegetables."[6]

The settlements in Gush Katif were part of a broader Jewish settlement movement in the areas that are colloquially called "Yesha" (a Hebrew acronym of Judea, Samaria, and the Gaza Strip; it also means "salvation").

Land appropriation and a struggle against territorial compromise have been at the center of this movement.[7] Whether inadvertently or deliberately, organic (and conventional) farmers took part in these efforts. In this chapter, I explore the conditions and discourses that led to the Yesha Jewish organic agricultural emergence. I ask how organic agriculture—which is about "sustaining the health of soils, ecosystems, and people, relying on adaptation to tradition and local conditions, as well as promoting fair relationships and good quality of life for all involved"[8]—can be produced amidst the conditions of colonization, exclusion, and partition? How can organic schemes be implemented according to a sociocultural logic of discriminative ethno-nationalism and ethno-religiosity?

I will show how organic Jewish agriculture in the occupied territories developed in three stages. At first, organic Jewish agriculture emerged on the basis of instrumental-scientific discourse, namely an agricultural discourse based on positivist "scientific knowledge," an allegedly non-ideological knowledge that embodies its own values of prediction and control.[9] Later, this scientific discourse was combined with more explicit ideological Zionist-national meanings (similar to those expressed by Mario Levi and described in Chapter 1). During the second stage, organic Jewish agriculture in the occupied territories evolved in accordance with discourse and practices that embody parochial and religious values and meanings. This spiritual-religious discourse created associative connotations between the meanings of organic agriculture and the ideology of the Jewish settlement in these areas.[10] From the pragmatic level, organic agriculture has been used as a mechanism for landholdings and for the establishment of communities of Jewish settlements in the occupied territories. Finally, after establishing the project of organic farming beyond the Green Line, Jewish settlers developed the organic agricultural enterprise according to globalized-neoliberal logic, emphasizing their connection to middle-class consumerism inside the Green Line and to the global organic market. In considering these three stages, I argue that Jewish organic farmers in these areas contributed to the construction of institutional, spatial, and cultural delineations between themselves and the Palestinian agrarian system.

## The Origin of Organic Agriculture in the Occupied Territories: Science and Zionism

The interlacing of organic agriculture, fundamentalism, religiosity, and political right-leaning is anything but new—and certainly not unique to the Israeli case. Looking at the political affiliations of organic agriculture from historical and global perspectives, one can find that many of the early practitioners and advocates of organic farming were associated with an anti-liberal worldview affiliated with conservative groups and even with extreme right-wing political movements.[11] In his book *The Global History of Organic Farming*,[12] Gregory Barton traced the ways in which the origin of organic farming was shaped by a cluster of ideas and processes, some of which are clearly not liberal: the tradition of yeoman independence in England, the British settlement colonies, the oriental-ist hunt for hidden wisdom in the East, neo-traditionalism, and even fascism with its emphasis on the vital link between culture and soil.[13] Examining the origins of the organic movement in the US, scholar Laura Sayre suggests that political and social conservatives have long formed an important element within the organic movement's ranks, particularly in the politics of organic farmers. According to her, histories of the organic movement should be told not only as histories of progressive or revolutionary campaigns to challenge the social order's status quo but also as "histories of conservative, even reactionary campaigns to restore a 'traditional' countryside."[14] The case of Jewish organic agriculture in Gaza Strip and the West Bank adds to this "global history of organic farming" as it reflects on the dialectics between fundamental religiosity, colonial settlement movement, and nationalism, on the one hand, and "green culture" and global consumerism, on the other hand.

Organic agriculture in Gaza and the West Bank emerged as part of the neo-Zionist project of establishing Jewish agriculture in the so-called "Greater Israel." In general, agriculture has served as a central means by which the settlement movement consolidates control over large areas of the occupied territories. Look, for example, at the West Bank: in 2013, over 93,000 dunams (22,980 acres) were designated for Jewish agriculture. This area was much larger than the actual built-up area of the settlements and outposts, which was about 60,000 dunams (14,826 acres, not including the Israeli neighborhoods in East Jerusalem). In this regard, the last decades have seen a decline of about one third in culti-vated Palestinian agricultural lands in the West Bank.[15] According to a report published by Kerem Navot (an NGO that deals with monitoring Israeli land policy in the West Bank), since 1997 settlers have taken over about 24,000 dunams of land through agricultural activity, of which about 10,000 dunams are on privately owned Palestinian land, mostly around the settlements and outposts in the West Bank. These agricul-tural activities can be seen as motivated by three interrelated reasons:

ideological religiousness, economics, and territorialism.[16] Though organic agriculture had been playing only a minor role in this broader project, it had been, and still is, of great symbolic and discursive significance.

The most significant role in the inclusion of organic agriculture in the movement of fostering Jewish agriculture in the occupied territories was played by Mario Levi, "the pioneer of organic agriculture in Israel," who was discussed in the previous chapter. Ultimately, he was also the first to bring the spirit, ideology, and settling practices of the Gush Emunim ("Block of the Faithful") movement to the Israeli field of organic agriculture.[17]

In 1983, while serving as the CEO of the IBOAA, Levi began to promote the project. He connected with farmers in Netzer Hazani, a settlement where the main source of livelihood was agriculture, and soon they established the first organic greenhouse beyond the Green Line. Later, organic greenhouses were established in other settlements in Gush Katif (among them Gan Or, Gadid, Ganei Tal, Atzmona, and Netzarim). In 2005, on the eve of the evacuation of Gush Katif, 700 dunams (173 acres) out of the 4,000 to 4,500 dunams (990–1,112 acres) of Gush Katif land were covered with organic greenhouses owned by Jewish farmers. The sum total of organic exports from Israel (including the occupied territories) reached US$55 million in 2004, of which US$10–12 million are estimated to have been from Gush Katif.[18] "[We had] great produce, produce with a high demand, very high quality, a quality that met every international standard," said Yossi, who was a resident of the area and chairman of the agricultural committee.[19]

The transformation of Gush Katif into a hub of organic agriculture and a major force in the Israeli agro-economy was accompanied by several discursive frameworks, most of which included agricultural-science and economic notions. When I asked Levi about the early days of organic agriculture in Gush Katif, he told me about the justifications he used at the time to obtain legitimacy for the project. Ideological justifications were not at the forefront of his actions—at least not on the rhetorical level. He argued that the development of organic agriculture was particularly suitable for these "uncontaminated" lands:

> [The lands in Gush Katif prior to the Jewish settlement] were cultivated in primitive ways. Many of these lands were "virgin soils" [non-fertilized cultivated soil], these were lands that were not cultivated using conventional methods, do you understand? It is first and foremost an agricultural matter, it

is a scientific matter, and these were the advantages and the qualities of the lands there.

According to Levi (and also drawing on some of the statements made by other senior organic farmers with whom I spoke), the lands of Gush Katif had not been contaminated with synthetic fertilizers, herbicides, or pesticides used in conventional agricultural methods; therefore, they were considered ideal for organic farming.

Another aspect in Levi's narrative on the origins of organic agriculture in Gush Katif is the hard work of "high-quality people" (*anashim aihotiyim*), namely hard-working and talented farmers, or, in his own words,

> Gush Katif was a center [of organic agriculture]. In the last year [before the evacuation] almost all of the agriculture there was based on organic greenhouses. Can you imagine that?! Gush Katif was one of the first places we started our activity. And it was an extraordinary success, very much because there were quality people there. They were successful farmers and succeeded in their work. Even the ordinary farmers [farmers from Gush Katif, who did not immediately practice organic farming] saw that it worked out well, so they learned from [the organic farmers]. It was "contagious." One [farmer] was "infected" by the other, because they were smart and saw that they can benefit from organic agriculture. They saw that it might give a little less crop, but at better prices. They invested in it and it was a huge success. After the evacuation [2005], many of them continued to work in organic greenhouses [in their new settlements], including those who were not organic farmers, because that was the specialty of Gush Katif.

He also mentioned that

> There were many Anglo-Saxons [in Gush Katif and other Jewish settlements in the West Bank],[20] and that was very important, because in these countries there is much more awareness [of issues such as organic agriculture, environmentalism, etc.]. The immigration [*aliyah*] from these countries, and the fact that they settled there, contributed enormously to the development of organic agriculture in Gush Katif, but not only there, also in Efrat and in other places in Gush Etzion.

However, Jewish organic agriculture in the occupied territories was developed not only due to rational, instrumental, or scientific-agricultural reasons but also due to the reinstatement of symbolic nationalism. Science and technology studies (STS) scholar Sheila Jasanoff argued that "policies of life science"—including techno-environmental debates on food and agricultural issues—may be "embroiled . . . in self-conscious projects of nation-building or, more accurately, in projects of reimagining nationhood."[21] According to Jasanoff, scientific knowledge—including the agro-science discussed by Levi and his colleagues—is grounded in political economic structures such as the nation state. Furthermore, identities and social order are always at play with science, and scientific discourses embed and produce themselves within society.[22] Jasanoff's arguments are well exemplified in the words of several farmers who said that the establishment of the organic sector in the occupied territories conveyed neo-Zionism (neo-nationalist ideology)[23] and visions of "the redemption of the land" in addition to the conditions and the agronomical qualities of the lands. For example, Erez, an organic farmer from Rechelim (a religious community in the hills of Samaria), claimed, "We determine the borders of the state of Israel through our organic agriculture. We believe this is the only thing that will allow this country to exist."[24] Similarly, Shivi, who served as a coordinator in the agricultural council for the central and northern West Bank, said, "It's difficult [to be an organic farmer in these places]. You have to be a staunch ideologue—not only regarding this [organic] growth method and lifestyle but also regarding Eretz Israel."[25]

Along these Jewish-ethno-national symbolic lines, the following saying was presented in the course "Basics of Organic Agriculture" (a course everyone who wishes to work as an organic farmer in Israel must take): "About 50,000 hens lay their eggs in the Galilee and in the Negev, and even on the hills of Nablus, Hebrew hens lay organic eggs that help to renew the historical continuity between us, the Jewish people, and the land of our ancestors."[26] In this regard, though emphasizing instrumental aspects, Mario Levi also told me that the development of organic agriculture in the West Bank and Gaza was inseparable from the ideologies associated with the settlement movement:

> After all, it was all about the connections between agriculture and the Jewish settlement in the Land of Israel. . . . It was not just about good lands and economic opportunities. They [the Jewish organic farmers who settled in the West Bank

and the Gaza Strip] were very devoted people, ideological people. . . . A while ago I visited Gush Etzion [a cluster of Jewish settlements located in the Judaean Mountains, directly south of Jerusalem and Bethlehem in the West Bank]. Fifteen years before I started to advance organic agriculture [with the farmers from Gush Etzion] they raised some apples and cherries there. The open fields were very few. I first arrived [in Gush Etzion] the week it was established. There was nothing there! And since then much has changed. I remember the person who pushed Kfar Etzion forward [in the agricultural and economic sense]. He built the foundations of the cow-shed, and I was there, working with him! This was a parallel move—agricultural success and a successful process of settle-ment. We got into organic agriculture with them, because we [the IBOAA] wanted to be part of the Zionist movement of settlement. . . . I've told you before: organic farming has always been connected to something bigger, to something spiritual.

Over time, organic agriculture in these areas began to seep into the religious spiritual discourse of more radical-Messianic settlers and shifted from the so-called bloc settlements (a cluster of Jewish settlements, such as Gush Katif or Gush Etzion) to the isolated outposts in the West Bank.

## Organic Food, Spiritual-Parochial Religiosity, and Territory

Mount Sinai Farm is an isolated outpost located on the outskirts of the settlements Susiya and Mitzpe Yair in the south Hebron hills of the West Bank (a region which is also known as Masafer Yatta). The legal status of the farm, as well as that of the neighboring Jewish communities, is controversial. It should be noted that, according to international law, all settlements in the West Bank are illegal. Yet, according to Israeli law, settlements built on state lands (officially confiscated or legally purchased) are recognized and legitimized. Illegal settlements and outposts are built on privately owned Palestinian land without authorization (but often with governmental support and assistance).[27] It was precisely on such lands that the founders of Mount Sinai Farm, Dalia and her late husband Yair, decided to "create roots in the land [of Israel] and spiritual roots in heaven," as written on the farm's website.[28]

Raised in secular moshavim (corporative agricultural communities) in the Central District of Israel, Dalia and Yair had strong connections to land and nature but not necessarily to religious-spirituality. They lived a comfortable farm life in the Hefer Valley (in a secular agricultural settlement in the Sharon plain in central Israel), but over the years they had doubts about their lifestyle: "This is a process that we started in the 1970s, after the Six-Day War [1967]. We felt that people were undergoing a process of materialism, competitiveness, and individualism—in contrast to the ideals and values on which the state [of Israel] was founded," Dalia recalled.[29] Concomitantly, they were exposed to veganism and adopted it as a way of life. They met Mario Levi while looking for opportunities to work in agriculture. Levi taught them about organic agriculture and suggested they travel to the United States if they wished to deepen their knowledge in agronomy and learn about organic farms there—and so they did.

However, their relationship with Levi led them to more than just organic agriculture. "Thanks to Mario Levi, the founding father of organic agriculture in Israel, we discovered our Jewish roots," writes Dalia on her farm's website.[30] In an article that covered her personal story, she says, "Mario literally exposed us to the importance of Sabbath observance, as well as to Jewish holidays and customs."[31] Later, Dalia and Yair attended seminars at Merkaz HaRav—a renowned national-religious yeshiva (Jewish educational institution) founded in 1924 by Rabbi Abraham-Itzhak HaCohen Kook. Merkaz HaRav has served as an important hub for proliferating the theology and ideology of national religiosity and has played an important role in the settlement movement in the history of modern Israel. In his writing, Rabbi Kook emphasized the sacredness of the entire Land of Israel. Zionism, according to him, is tangible proof of the divine decision to return the people of Israel to their land, and proof that the borders of "The Promised Land" should be determined through halakhic deliberation (Jewish religious laws derived from the written and oral Torah) rather than historical contingency or diplomatic negotiation.[32] "These were the years of our personal and spiritual molding," said Dalia.[33] She and Yair moved to the Susiya settlement, a Jewish settlement that was established in 1983 near the Palestinian village of Khirbet Susiya, on Palestinian land that was declared state land by Israel. Thus, while they were in the process of returning to the faith (*hazara betshuva*), they established their farm. Avidan, who later served as the director of the farm, said, "When Yair (the co-founder of the farm) started a process of turning to God, he also began a process of returning to our sources and the true culture we want to bring back to this land. He led a movement in this area of a healthy and genuine connection to the soil, and not of people who live in the Land of Israel [but adopt] Western culture. He set up this special sheep farm [as a means of] being connected to the Land of Israel." In fact, Yair learned from his neighboring Arabs about shepherding and imitated them.

A famous song by Israeli poet and singer Meir Ariel, "Shir Ke'ev," refers to the early Zionist practices of appropriating Arab attire and manners: "At the end of each sentence you say in Hebrew sits an Arab smoking his *nargila* [hookah, an Arabic water pipe also known as *shisha*]." In the early days of Zionism and Jewish settlement in Palestine, the image of the Arab farmers (the *fellahim* in Arabic) gained a mythological status and served as a role model to the Jewish settlers due to their attachment to the land.[34] Dalia and Yair followed this approach and adopted dryland farming (*baal*).[35] "We learned to rely on the rain from the sky. [All we know and do] we learned from the Arab *fellahim*," Dalia said.[36] She also added,

> Yair learned from Arab shepherds and realized that the way to protect the land would be to raise sheep. . . . We were the first [Jewish shepherds] in Judea and Samaria. At first the [Jewish] residents of the settlement [Susiya] were surprised and told us "shepherding is only for the Arabs. Why are Jews raising goats and sheep?" But Yair did not give in and we bought a small herd. Our relationship with the Arabs was a relationship of mutual respect but also of mutual suspicion. At first it was very flattering to them that Yair followed them. He even learned to speak Arabic fluently. But then they saw that [through herding] the land returned to us, and they were less enthusiastic. . . . Yair went out every day with the herd to graze around Susiya, and slowly the Arabs withdrew.[37]

The boundaries between Dalia, Yair, and their neighbors, the Arab *fellahim*, were delineated not only at the spatial level but also in the symbolic and metaphorical levels. Outwardly the farm and its organic agricultural practices were never presented as imitations or adoptions of Arab agriculture in the region, but rather as a symbol of Jewish tradition and continuity. When describing the establishment of the farm, Dalia avoided the use of terms such as *baal*. Instead she insisted on pouring biblical meanings to their farming practices:

> Our agriculture enabled us to connect the corporeal with the spiritual. In the Southern Hebron hills, we were exposed to traditional agriculture, agriculture that exists from the rain of the sky and consists of four elements: cereal grain crops [*dagan*], wine [*tirosh*], oil [*yitzhar*], and cattle [*behemah*]. [Quoting from the bible:] "The land, whither ye go over to possess it, is a land of hills and valleys, and drinketh water as the rain of heaven cometh down"; "that thou mayest gather in thy corn, and thy wine, and thine oil. And I will give grass in thy fields for thy cattle, and thou shalt eat and be satisfied" [Deuteronomy 11:15]. So, in addition to a herd of goats and sheep, we sowed fields of wheat and barley, planted olive groves and vineyards. . . . The idea was to achieve an end product from each of these elements. To bake bread, to make cheese from our own dairy. Yair also built a small winery, but we realized that making wine requires a lot of skills, so we produced *tirosh*, an organic grape juice. It was delicious.[38]

Currently, Jewish-biblical symbolism (similar to the one expressed by Dalia) is prominent among many of the farms and food production plants in Jewish outposts and settlements in the West Bank. For example, Ben's organic winery, located on the outskirts of the West Bank settlement Yitzhar, produces wine named Shoham after the stone in the sacred priestly breastplate (*hosen*) worn by the High Priest of the Israelites according to the Book of Exodus. One of the wines is called "Pura" after one of the words for a winepress used in Isaiah. Ben, the owner of the winery and a former yeshiva student, wrote the texts used on his wines' labels. One reads, for example, "The grapevines put down roots in the earth of the good land, and breathe its air until the roots strike the heart of the land."

Jewish-biblical symbols, such as those used by Dalia or Ben, are often accompanied by other structural aspects: the emphasis on Jewish or Hebrew labor (*avoda ivrit*), namely the emphasis on hiring Jewish rather than non-Jewish workers, as well as physical actions of pushing out Palestinians and the use of agriculture to create borders and territorial boundaries. For instance, the label of Ben's Pura wine says, "The wine production process, from the planting of the grapevines to the sealing of the bottles, uses *Hebrew labor* of the young people of Israel, to give pleasure to the Creator, may His name be blessed." In this same vein, Dalia from Mount Sinai Farm says, "We grow barley, and we buy organic Israeli wheat from Sde Eliyahu. We use other kinds of flour in our mill such as organic spelt flour and organic rye flour. . . . All is done here on millstones. We use *pure Hebrew labor*—from beginning to end."[39]

The emphasis on "Hebrew labor" is often accompanied by other symbols that emphasize the culture of being connected to the Land of Israel (*Eretz-Israeliyut*) and Hebrew culture, which, nowadays, is fostered by religious-Zionist groups (as opposed to Israeli culture [*Israeliyut*], which is associated with liberal sectors in Israeli society that do not necessarily emphasize their connection to the land). For example, the text on the labels of organic olive oil produced by Achiya Farm, located in Shilo (a Jewish settlement in the northern West Bank—"Israel's Biblical heart-land," according to Achiya Farm's owners[40]) reads, "The [olive] groves that characterize the landscape of the Land of Israel symbolize a rooted and traditional history. We continue to maintain the Hebrew tradition [of olive growing and olive oil production] in the Land of Israel. We cultivate, with love and devotion, hundreds of acres of olive trees of various species and promote modern Hebrew labor. Organic olive oil consists of cultivars grown in our olive grove plantation." Thus, the Achiya Farm claims a historic connection to oil production despite the fact that, historically, olive oil was rarely produced by Jews in modern Israel, and that, until the 1990s, olive cultivation was a typical Palestinian-Arab occupation.[41]

Furthermore, actions of "pushing out the Arabs" are also common. As Dalia mentioned, "After 17 years, we have accomplished an achievement which is unparalleled in Judea and Samaria: following the withdrawal of the Arabs from the area, [we cultivate] 10,000 dunams of pasturage and crops." Avidan, the manager of the farm, testified that "Yair used to go out to the open spaces. He initiated a new spirit of true connection to the Land of Israel and to religion. One day when he went out with his herd, he was murdered by terrorists . . . but thanks to him, and to the grazing lands he took over [from the Arabs], the Susiya settlement has a sort of security belt of a few kilometers."[42]

These vignettes exemplify how organic farming, labeling, and marketing are practiced through the construction of a symbolic "regime of authenticity," a "taste of the Land," as well as temporal continuity of Jewish tradition and connection to the land[43] (i.e., attaching biblical names to organic wines, labeling olive oil "from Israel's Biblical heartland," etc.). From a structural perspective, organic agricultural practices serve as a mechanism for control over land and labor, and thus promote the realization of a "split labor market" in these areas, to use the terminology of sociologist Edna Bonacich[44]—namely, a set of segmentations on the basis of race (or ethno-ethnicity in the Israeli-Palestinian case).[45] All these symbolic and structural practices are done through rescaling the macro meanings of "organic" into the specific local politics of this contested region.

## The Food on the Hill:
## Organic Farming on the West Bank Hilltops

A reporter from the Israeli religious and nationalist newspaper *Makor Rishon* covered the story of Dalia and Mount Sinai Farm. He stated, "The farm is quite small, but the gospel of the farm made a huge impact. . . . They initiated the concept that the renewed Jewish settlement in Judea and Samaria should also be agricultural. . . . Over the years there were many moments of depression in this place, but nowadays Jewish agriculture in Judea and Samaria is already a known fact! . . . [Mount Sinai Farm] was actually the first 'hill' in Judea and Samaria, long before anyone referred to the term 'youth of the hills.'"[46] As scholar Michael Feige explained, the term "youth of the hills" refers to the young settlers who hold a radical rightist standpoint. The ideal of settling the land and resisting any attempt at territorial compromise is paramount to their identity. These young settlers try to expand the Jewish settlement by leaving the established villages and towns in the occupied territories and capturing a hill somewhere in Judea or Samaria in clear violation of Israeli law. Many of them are highly motivated to "work the land, to harvest olives . . . , to show who the land-lord around here is, not to hide inside the settlement but to step out."[47] Ultimately, this new settlement movement involves practices of land-grabbing.

"The revolution arose from our farm," Dalia said, referring to this trend of private settlements and the initiation of outposts. Her use of the word "revolution" is akin to the way in which Michael Feige describes those young settlers as "the flower children of the hills."[48] Sociologist Shlomo Fischer describes the emergence of radical religious Zionism and its manifestation in the activities of "the youth of the hills" as a transition from collectivism to individualism within the settlement movement.[49] According to Fischer, [the "youth of the hills"] foster the radical religious Zionist theme of imbuing the land (i.e., the Land of Israel) with religious meaning and value . . . emphasizing its immediate-ness and concreteness. The outposts and the hilltop settle-ments . . . are spread horizontally over distances so that each farmhouse merges with the rocks, soil and olive trees in its immediate vicinity. They make a point of wanting to be on the land in the most unmediated way; living in direct contact with the elements. Many of them practice organic farming, they tend to wear homespun clothes, noteworthy for their simplicity, and build their own homes from materials locally available. Considerable numbers of them eschew electrical sources of energy, using wind or solar power. . . . Thus, they interweave the themes of creative arts, counter-cultural and New Age ambience, ecology and unmediated contact with nature, the earth, its elements and violence.[50]

In this regard, Dalia insisted that she and her late husband Yair laid the foundations of what she describes as the "back to nature" culture, which is now common among the second generation of Jewish settlers in the West Bank. "We were ahead of our time on many levels," said Dalia. "We talked much earlier about things that people only now under-stand. . . ."[51] Almost every [Jewish] shepherd you'll talk to in Judea and Samaria worked here," she says. "Even Avri Ran and his wife Sharona. She made her first *labneh* here.[52]

The person Dalia mentioned, Avri Ran, was dubbed by Feige as "the father figure of the outposts."[53] Journalist Aviv Lavie[54] called him "the sheriff of the hills."[55] Just like Yair and Dalia from Mount Sinai Farm, Ran grew up in a secular community—in a kibbutz—became religious, and together with his wife Sharona moved to the settlement of Itamar in Samaria. Soon after, in order to "strengthen the ties with the land, and [to] draw closer to God,"[56] he decided to move to the top of one of the high hills, east of the settlement Itamar, and set up his outpost. The outpost and farm he established were named Givo't Olam (meaning "the hills of the world" or "the everlasting hills"). The name of the farm, which was suggested by his wife Sharona, embodies various symbolic meanings. It is taken from *v'zot ha'berachah* ("and this is the blessing" in Hebrew), a portion of the Torah (*parashah*) in which Moses called on God to bless the tribe of Joseph: "Blessed of the LORD be his land; for the precious things of heaven, for the dew, and for the deep that coucheth beneath, And for the precious things of the fruits of the sun, and for the precious things of the yield of the moons, And for the tops of the ancient mountains, and for the precious things of the *everlasting hills*" (Deuteronomy 33:13–17).[57] According to the biblical narrative, the two half-tribes of Joseph settled in the same area where the Givo't Olam farm is currently located. "We must connect the mountain communities of Samaria to the Jordan Valley," said Sharona. "This is an indispensable corridor of settlement."[58] According to journalist Aviv Lavie's interpretation, the organic dairy products and the vegetables Ran and Sharona produce are embedded with symbolism of both health (*briut* in Hebrew) and creation (*briah*, a word that is phonetically similar to *briut*).

Just like his colleagues, the other Jewish farmers and settlers in the outposts of the West Bank, Ran insists that "everything on this hilltop was built using *avoda ivrit*—Hebrew (Jewish) labor. When he [Ran] built his first chicken coop in Itamar he hired young settlers and gradually became not only their boss, but a father figure."[59] Gradually, he became "an icon of the radical right, a local organic farming guru . . . and the founder of the narrative of the outposts," as defined by one of his neighbors. "Ran embodied and manifested—in and of himself, through his way of life, through his own body and through his work—the realization of the triad of man-land-God."[60] Many of those dubbed the "hilltop youth" gathered around Ran's ideology of Jewish self-sacrifice for the Land and Nation of Israel. They followed his path and engaged in the hard work of organic farming in those remote hills. A reporter for *Arutz Sheva*, an Israeli media network aligned with religious Zionism, praised him: "Many others have emulated Avri's way of settling the land—leaving the confines of the gated communities for the biblical bounty of the barren hills and valleys of Judea and Samaria and making them blossom."[61]

It is hard to think of more dramatic descriptions than Ran's, when he talks about his connection to these hills: "Every egg I collect, every chicken I care for, and every animal or grass in the yard is part of my covenant with the Land of Israel and the Torah of Israel. . . . This land is my own self and my own flesh, it is my very existence. I breathe the air from above and suck the earth from below as well. If you ask whether I will sacrifice my life for this place—my answer is yes, certainly yes, I will sacrifice my life for this place, absolutely. Because this place might require sacrifices."[62]

It should be noted that contrary to this aggressive and apocalyptic terminology ("sacrifice my life"), the farm is often portrayed as aesthetic and picturesque, such as in the following quote taken from a newspaper article about the farm. "In a large goat shed," writes Yair Lapid, a former journalist and current politician, "between the organic goats, there is a wooden platform with a piano on it. During milking, one of the farm girls plays for the goats, so that their organic milk will come out tasting of Mozart."[63] Additionally, a visit by the members of the IBOAA to the farm was summarized as follows: "Givo't Olam looks like a wild utopian farm: [it is surrounded by] clear mountain air, green lawns and olive trees. During the tour of the farm, we even saw two *bambis* run past us"[64] (the writer refers to the well-known deer of the famous book *Bambi: A Life in the Woods* and the Disney animated film in an attempt to simulate an image of the farm as a magical place).

However, the reality is far from the pastoral picture that emerges from these descriptions. For the nearby Palestinian village of Yanon, Ran's outpost spelled disaster.[65] As political scientist Hagar Kotef describes, Givo't Olam was founded by means of land-grabbing, intimidation, harassment, and physical violence. She specifies that Ran's dedication to organic farming is consistent with his worldview. According to him, brutality and aggressive acts are part and parcel of being in nature, cultivating lands, and practicing agriculture. Organic agriculture, to him, is no exception, and thus serves as a mechanism for justifying his actions and of reconciliation with the reality of violence that surrounds his outpost and farm.[66] Indeed, Ran's conduct has even reached the Israeli courts, where he was indicted several times for assault under aggravated circumstances.[67] Michael Feige argued that this case is typical of cases related to the "youth of the hills," including the intense hostility toward the Arabs.[68]

In this regard, I found it striking that, despite the fact that these settlers are preoccupied with the search for agricultural authenticity and connection to the land (dryland farming and shepherding, for example),[69] they still distinguish themselves from Palestinian agriculture, which is compatible with the local environmental conditions and might even be considered authentic, organic, and respectful. For example, Erez, an organic farmer from Rechelim (a religious community in the hills of Samaria), claims, "In my naiveté I thought the Arabs [Palestinians] farm organically, just like 3,000 years ago. But . . . the Arabs in our area use substances that are no longer up to standard."[70] Similarly, Ran described the Palestinian agriculture not as natural, sustainable, or organic, but as primitive. "They don't have the motivation to be good farmers. It doesn't interest them. Seeing them plowing the soil with a donkey and harvesting olives by hand from the trees—that isn't even agriculture—organic or otherwise."[71]

In point of fact, fair trade cooperatives have helped 3,000 Palestinian agriculturalists in the West Bank receive fair trade and organic certifications and sell high-quality produce in international markets.[72] Further, it is noteworthy that within the borders of Israel, there are only a few Palestinian farmers (citizens of Israel) who are certified as organic.[73] Thus, to paraphrase Julie Guthman's words, which pointed to the "unbearable whiteness of alternative food initiatives in the United States,"[74] it might be fair to say that organic agriculture inside the Green Line has developed to be unbearably Jewish: distinct, culturally and practically, from Palestinian agriculture, not to say hostile to it.

## Neoliberalizing the Organic Outposts: From the West Bank to Tel Aviv

Dalia, the organic settler farmer who currently runs Mount Sinai Farm, once announced that she strives to create "paths between materiality and spiritualism": "I hope that people in Israel will be able to recover from the total attachment to Western culture and to the culture of abundance that so disrupts our lives."[75] That statement may be consistent with the common sociological account about the lifestyle and ideology of the so-called "youth of the hills." This account points to the opposition of this young generation of Jewish settlers against the lifestyle of their parents. The latter, religious Zionists associated with the Gush Emunim settlement movement, were characterized by the modern-bourgeoisie lifestyle. Ran, Dalia, and their followers might instead be seen as those who chose an alternative, antinomian, and avant-garde lifestyle.[76] And in many ideological and political aspects, they did. But if the picture that emerges is that of farmers who work on small properties to provide for their modest needs while eschewing "Western culture" or modern-instrumental practices, it is certainly far from reality.

In Givo't Olam, for example, one can find 70,000 free-range, organically raised chickens. Ran and his workers also produce the food for the chickens (a mixture of organic clover and oats that grows in the 450 dunams of lands around the outpost, which is prepared in a feed mill Ran built). They run a packing house and work on marketing and distributing their organic eggs within Israel and the Israeli settlements in the West Bank. Nowadays, Ran's farm is the largest farm in Israel and Palestine that produces organic eggs. From the establishment of his farm to today, there have been between 15 and 25 certified organic poultry farmers in Israel and Palestine. It is estimated that Ran produces more organic eggs than all the others combined. Israel's poultry farmers' association considers him a "big producer," although he produces only organic eggs. To put things in perspective, all the organic eggs produced in Israel do not exceed 2% of the total amount of eggs produced in Israel; Ran, who produces only organic eggs, is also one of the largest distributors of all eggs in Israel, regardless of production method (organic and non-organic).[77]

In addition to the organic egg business, Ran built a goat pen, raises thousands of goats, runs a dairy, and produces a variety of organic dairy products. Currently, the vast area that Givo't Olam covers includes animal pens, olive groves, fields of organic vegetables, a flour mill, loading and unloading platforms for customers and suppliers, a dairy, a vineyard, offices for marketing, and a synagogue. Ran, who used to boast about being a "self-made man," insists that he built this impressive organic farm literally with his own hands, without any support from any organizations or governmental or regional authorities. It should be noted, however,

Figure 2.1. A Givo't Olam truck parked in the parking lot of a branch of a conventional supermarket chain at the Gush Etzion Junction. Photo by Dror Etkes. Courtesy of Kerem Navot.

that while the legality of the farm is contested (or holds a legal status that only few outside of Israel would be ready to grant), Ran is supervised and receives organic certification from Agrior. Currently, Agrior is a private organic certification and inspection agency, but between 1999 and 2011 the company served as a subcontractor of MARD. In addition, Ran receives governmental veterinary supervision and works under a quota of eggs determined by the Ministry of Agriculture.[78] Thus, due to the state's direct and indirect support of Ran's production and marketing processes, and the growing demand for the farm's organic produce, one can find Givo't Olam's produce in numerous stores throughout Israel.

A CEO of a large organic distributer, who testified that he is not a right-wing political supporter and thus might not agree with Ran's practices, nevertheless buys and distributes Ran's produce. He said, "We are not a political company. We are an organic company."[79] The owner of another retail chain that sells organic produce said, "All our organic produce is legally certified. Whoever objects to the produce coming from settlements and outposts, I suggest that he stop eating at all. Does he buy coffee? Chocolate? Clothes from China? I visited factories in China, and I couldn't believe what I saw there. You say that Ran's farm is an illegal outpost? Well, to the Muslims, we are illegal in all of Israel. I'm not a right-wing supporter, but I will not support banning any of the produce from the settlements." Similarly, the marketing director of a large organic retail store from the Central District of Israel said, "We do not deal with political issues, we have customers who hold all kinds of political opinions—and that's how it will remain. . . . What matters to us are the organic standards."[80] Ornit Raz, who served as the director of the IBOAA, said that the association "doesn't deal with politics, such as where the organic farms are located. We examine the produce that comes from everywhere, according to its quality and our professional standards."[81] And a CEO of a marketing company that distributes Givo't Olam's produce throughout the country said, "This is the most beautiful organic farm in the country, and its products are the best!"[82]

This emphasis on the high quality of Ran's produce is compatible with the broader progressive politics of mediating foods and other commodities produced by Jewish settlers in the West Bank; these new strategies aim to divert attention from the colonial condition within which the foods are produced.[83] By emphasizing the culinary value of the food produced in Givo't Olam, Mount Sinai Farm, and the like, organic farmer-settlers place their produce in the center of the field of organic food in Israel, and sometimes even in the international organic field. These strategies may be termed economic and commercial *gastro-diplomacy*, namely, the use of foodways (food production and trade, in this case) to create, maintain, or enhance a country or region's brand.[84] Gastrodiplomatic tactics are applied by Ran and his colleagues (and comrades) to foster the commodification of settler organic agriculture and thus promote the normalization of their existence—all while hiding their contested/colonial conditions.

In keeping with marketing strategies adopted by other food producers in the West Bank, such as winemakers, organic food producers embodied the discourse of "YESHA is Fun" (*YESHA ze fun*), namely, a cultural-political colloquial method of convincing the Israeli public that the settlers are not messianic fanatics but ordinary Israelis, and that the places in the West Bank beyond the Green Line, such as Susiya or Itamar, and even the hilltop outposts, are no different from places within the Green Line.[85] The guidebook *YESHA is Fun: The Good Life Guide to Judea and Samaria* (2011)[86] reads,

> The Mount Sinai Farm, which is located next to the houses of the settlement of Susiya, is an ecological-organic-family farm. . . . At the entrance, one can see the home-style dairy. . . . [When you visit there] in the heat of the summer, smiling people will give you cold sweet water from the house's water well, and tastings of fine sheep cheese. Sounds like a dream? Like a hallucination? We do not know what they put in their cheese, but it really felt like a dream. Buying cheese from Mount Sinai Farm is recommended for connoisseurs. Some of the cheeses are salty, others—refined in their taste. But everything has a homey, simple and non-commercial taste. Not only dairy products can be found at Mount Sinai Farm. There is also flavorful organic flour, whose health benefits are well known. The Mount Sinai Farm products can be found in Be'er Sheva, Jerusalem, Mevasseret, Efrat and of course, here, at the farm.[87]

The Mount Sinai Farm website mentions that they market their products in the south of the country, Tel Aviv, and other places in the center of the country. Indeed, organic produce produced beyond the Green Line is ubiquitous, and the Jewish organic settler-farmers seem to boast about it. Ran, for example, testifies for himself and for his brand: "Wherever there's a health store, our products are there. . . . Whether we're talking about products, business, integrity or ethics . . . [Givo't Olam is] considered the most professional organic farm in the country, in my opinion. Yesterday someone tried to arrange a meeting with me to set up an organic farm in Europe. At the organic level, we definitely have no competitors. We sell millions of eggs annually to the organic market—they're the best organic eggs out there. They reach thousands of households in Israel."[88]

About Ran, Ornit Raz, a former director of the IBOAA, said, "aside from being responsible for much of the growth of the organic egg industry, [he] also uses his thousands of dunams to grow wheat used to make organic flour, and produces cheese and yogurt made from goat's milk. He is an excellent farmer, straight and reliable. We, as an association, look at him from a professional and ethical [agricultural, environmental] standpoint, not a political one."[89] Ornit went on to say, "Givo't Olam is a respected brand in the professional clique, and among organic farmers and many health-food stores [consumers]." However, Ran is aware that there are individuals who wish to avoid his products as a matter of conscience, or, as he puts it, politics: "Some people mix food and politics. There are also consumers in Tel Aviv who look for our products," he says, referring to consumers from both ends of the political spectrum: those who do not care that the organic products are produced in the West Bank settlements and wish to buy Ran's products because of their quality, and those who wish to boycott them because they are seen as products of settler colonialism. "So, we have to deal with that," says Ran. "Sometimes, Givo't Olam products aren't sold under the brand name. That is how leftists, the most leftwing of the leftists, end up buying Givo't Olam products."[90]

The combined stories of organic farming in Gush Katif, Gush Etzion, Susiya, and the hilltops adjacent to Itamar illustrate the different phases and the changes through time of Jewish organic agriculture in the occupied territories. In the beginning, this agricultural project was justified by agro-scientific explanations as well as by claiming Jewish indigenous ownership of "Eretz Israel" and a spiritual connection to the land. Later, this organic-settlement project became quality-commercial oriented. This "quality turn" continues to foster the distinctions between the Jewish farmers-settlers and the Palestinian farmers. In addition, it bonds together radical religious neo-Zionist settlers from the hilltops and post-Zionist liberal and secular consumers ("the leftists" to use Ran's terminology)—a connection based on the production and consumption of high-quality organic products. The following two chapters deal with the latter: the post-Zionist producers, distributers, and consumers who engage with organic food not only as a means of cultivating a "lifestyle of health and sustainability (LOHAS),"[91] but also as a means of fostering a cosmopolitan identity.

Chapter 3

# Organic Start-Ups

## CSA, Farmers Markets, the Creative Class, and the Atmosphere of "Being Abroad"

Israeli author, poet, and journalist Yitzhak Laor argued that "abroad" (*chutz-laaretz*) is one of the most significant terms in contemporary Israeli culture. He substantiates this claim by pointing to the tremendous efforts made by middle-class Israelis to travel overseas or to just "be abroad" for any reason: vacations, studies, visits to relatives, business, and more.[1] Similar to Laor's claim, this chapter argues that the notion "abroad" is crucial to understanding forms of food production and consumption among middle-class and upper-middle-class groups in Israel as revealed through the study of organic food. As described below, expressions of "being abroad" are ubiquitous in the discourse of small-scale Israeli organic farmers and their consumers, and they are manifested in places that might be considered "local." This chapter elucidates how organic foods serve as global culinary artifacts and how small-scale local initiatives work as vibrant places for importing, implementing, practicing, and negotiating "global culture."

In the Global North, organic food is often related to the agri-cultural-culinary paradigm, which is based on interest in "alternative," "quality," and/or "local" food networks. Nowadays, this paradigm is mostly associated with initiatives such as farmers markets, small farm shops, community-supported agriculture (CSA), home deliveries, and box schemes—all, at least discursively, opposed to the industrial and conventional food system. Geographers Brian Ilbery and Damian Maye summed up this binary opposition in global public discourse as between what is often conceived as "alternative" agro-food and conventional food systems. They point out that terms such as *short food supply chains, embeddedness, traditional, regional palates,* and *local markets* characterize alternative food production systems, while terms such as *long food supply chains, global hypermarkets,* and *disembeddedness* characterize conventional food production systems.[2] A long list of critical studies in recent years, including the research Ilbery and Maye conducted, point out that these binary oppositions are never simple or clear-cut, and that the distinctions between conventional and alternative systems are often blurred.[3]

While most of these studies focus on the political and economic levels, this chapter explores the cultural levels that blur "conventional" and "alternative" foods. It focuses on CSA and farmers markets in Israel to show how these segments of the organic food field—which are often considered "local" initiatives operating in the form of "short food supply chains"[4]—are actually embedded with cosmopolitan similes and meanings. In addition, I argue that practices of organic food production and consumption in these initiatives play a part in the process of middle-class-making in Israel. Rather than acting as agricultural or agronomical hubs, CSA and farmers markets in Israel serve as a means of demonstrating cultural "openness to the world" and constructing an Israeli version of global habitus.[5]

## CSA in Israel

Community-supported agriculture is a model for growing, marketing, and consuming agricultural produce; it first appeared in the second half of the previous century in Switzerland, Germany, and Japan. Ideally, the model is based on a variety of partnerships and mutual support between farmers and consumers.[6] It is often realized when a group of consumers join a farmer and commit to regularly purchasing his or her crops. The farmer, in turn, ensures quality, reliability, and locality of produce. Rhetorically, many CSA programs use the notion "members" rather than "consumers." Ideally, members share the risk of production in exchange for a share of the produce. In practice, CSA members buy products directly from a farm, pay in advance, and the farmers do their best to produce sufficient quantity, quality, and variety to meet members' needs. Within the global field of organic food, which is conceived as divided between corporate-conventionalized organic and resistant-countercultural niche farming,[7] CSA is firmly associated with the latter. Since its inception, the CSA model has been perceived as ecologically standing in contrast to chemical and monocultural agro-practices.

The CSA model is often conceived as a model of "moral economy"[8] distinct from the economic structure of industrial agri-businesses. It is often described as a model that fosters the reestablishment of communal relationships to encourage environmental and social citizenship rather than self-focused consumerism.[9] Supporters of the CSA model describe it as a marker for the re-infusion of sociocultural aspects of working the land, as operating outside market rules, as a clear contrast to the process of commodification of agriculture, and as a real expression of opposition to globalized and industrialized agriculture. Critical accounts, particularly those that employ a Marxist political economic understanding of the social embeddedness of the CSA model, point to problems within this model, such as instability and differentiations in farmers' earnings and issues of farmers' self-exploitation derived from this model.[10] However, as geographer Ryan Galt noted, there are a variety of definitions of CSA, since the concept itself, and the practices that constitute it, have been substantially modified when applied to different locales.[11] What, therefore, are the affiliations, modifications, and social affinities within the CSA model as it has been assimilated in Israel? What are the sociocultural aspects upon which this model is contracted in Israel, and what are its sociological meanings?

Currently, around 20 farms operate according to the CSA model in Israel. All of them, without exception, grow certified organic produce. This situation is somewhat different from CSA in other places in the Global North where it is often certified organic or advertised as based on "beyond organic" practices—namely, mutual relations that are based on transparency and trust without the mechanism of third-party organic certification. In Israel, all CSA farms provide a weekly box of certified organic fruits, vegetables, and other organic foodstuffs to several hundred households.

The first to explicitly adopt the CSA model in Israel, and later influence many others, was Bat-Ami Sorek. Sorek returned to Israel in 2002 from a long stay in California, where she was first exposed to organic agriculture and community gardening. She studied organic agriculture and sustainable communities and holds an MBA from the New College of California. She also experienced working on several farms in California, where she was engaged with direct marketing from farmer to consumer. At that time, the popular discourse in the USA and other Global North countries favored "local," not only "organic," as a key solution to the ills of the industrial food system.[12] In line with the global discourse that held notions of "local food," Sorek began operating a farm and a CSA program in Israel, calling it Chubeza.

It is hard to think of a name that embodies more meanings of localness, rootedness, earthiness, and connections to the history of modern Israel than the word *chubeza*. Also known as *helmit* (in Hebrew) and *mallow* or *malva* (in English), *chubeza* is the Arabic word for a type of edible wild plant. During winter and spring, chubeza plants are widely spread in any open field. Although the plant is nutritious and abundant in the Middle East, it has never been cultivated by conventional farmers in Israel and is thus embedded with meanings of nature and wilderness. However, chubeza is one of the wild herbs at the center of traditional Palestinian Arab foodways.[13]

Chubeza is also associated with the word "bread," both in spoken Arabic and Hebrew (in Hebrew it is referred to as "Arabic bread"). Importantly, this plant is embedded with significant symbolic and historical meanings in the Israeli context. It holds a place of honor in the Zionist narrative and in nationalist popular discourse. In the midst of the 1947–1948 Civil War in Mandatory Palestine,[14] during the battle for Jerusalem and the siege of the city (November 1947 to July 1948), it was chubeza (malva) that helped improve the meager nutrition of the Jewish population. In naming her CSA farm Chubeza, Sorek conveys local-historical meanings, given the farm's location in the fields of Kfar Bin Nun on the Latrun–Ramle road in the Ayalon Valley on the way to Jerusalem. This place is known to have been blocked by Arabs in order to prevent access to Jerusalem ("Latrun and Bab al-Wad") during the 1948 Arab-Israeli War and the siege of Jerusalem. Thus, global ideas about the "local food movement" and agricultural models that were imported by Sorek from California were integrated with local-historical and national symbols shared by both the Jews and Palestinians who live in Israel and Palestine as epitomized by the name of her farm.

Chubeza, the farm, was established in 2003 on a small field. In 2009, the farm cultivated 50.5 dunam (12.5 acres) of farmlands, and gradually increased to 101 dunams (25 acres) in 2019. Currently, Sorek and her team cultivate a biologically diverse vegetable field with 100 different varieties of vegetables and herbs—all organically grown and certified. The demand for the farm's produce increased gradually: during the first week of its operation (2004), the farm produced 11 vegetable boxes. Two months later, the demand increased, and they produced 45 boxes. In 2008, they produced 350 boxes, and since 2011 Chubeza has been sending out more than 450 boxes of produce a week.

A few years after the establishment of the farm, Sorek received some media coverage, during which she introduced the CSA model into the public-culinary discourse in Israel. Soon, she was recognized as the farmer who "imported" the model to Israel, and the one who was responsible for its assimilation into the Israeli field of organic food. Following the success of the farm, the CSA model attracted many young farmers who were interested in alternative agriculture. Gradually, other farms operating according to Sorek's agricultural methods emerged. Beyond the agricultural patterns, these farmers also followed her cultural and symbolic path as they adopted the discursive frame that incorporates similes of both the global and the local.

The CSA model itself is portrayed in Israel as global, progressive, innovative, and up to date. For example, Sorek and her colleagues insist on referring to the model according to which their farms operate using the English acronym "CSA." They often lecture about the global history of CSA, openly present their awareness and concern for global environmental issues as well as global food justice issues, and are culturally fluent in the global discourse of alternative food movements. For example, Sorek wrote the following lines in one of the newsletters she sent to her consumers, the members of Chubeza Farm:

> CSA Today! . . . I'm devoting this newsletter to discussing the CSA, its history and the present state of the movement, which attempts to create partnership, responsibility and reciprocity between farmers and consumers. . . . This model arose simultaneously and independently in both Japan and Europe (in the pre-"like us on Facebook" era). This happened in the 1960s. . . . Countries were losing farming viability and the wherewithal to sustain independent farming in an era of global market economy and low-cost imports. In short, people began to be aware of agricultural problems. . . . Japan is a country with a longtime tradition of cooperatives. . . . In 1965, they integrated the *teikei* (Japanese for "cooperation" or "joint business") model into the effort. In reference to CSA, it is commonly translated as "food with the farmer's face on it." . . . Meanwhile, on the other side of the world . . . at the end of the 1960s, the Buschberghof Farm, a German collective farm based on these principles, was established alongside a

"collaborative agricultural community." This initiative aimed
to create a network of non-farmers who support farmers by
giving loans and partnering with them. In Switzerland, a
similar process took place. The development of the CSA
movement in the United States was quite similar to that of
its European counterparts.[15]

At the same time, Sorek ascribes local-national cultural meanings to the
CSA model. This is exemplified in her thesis project submitted to the
New College of California, in which she described the establishment of
Chubeza Farm:

> The first Zionist pioneers wanted to "redeem the land" by
> "taming the shrew," sometimes in a forceful manner, pouring
> cement and asphalt on the land and compacting it with
> giant tractor wheels. Still, there was a lot of love in this
> relationship; love of the land, love for the nation, love for
> agriculture. Building a CSA in Israel in this day and age is
> a renewed act of redemption of the land.[16]

However, these images and meanings—of a farmer motivated by love of
the land and love for agriculture—are significantly different from those
of the "Zionist Pioneers" to whom Sorek refers. While the latter's practices were conceived as part of a larger project for the establishment of
a national home for the Jewish people in Palestine, the former seem to
be engaged in post-national cultural discourses.

The sociological discourse about the middle classes of the late
twentieth century and the early days of the twenty-first century—the
time in which CSAs were established in Israel and proliferated in other
locations—was focused on the predominant cosmopolitan lifestyle and
wide range of the then-new possibilities of detaching from the "local"
and crossing real and imaginary boundaries.[17] Cultural globalization in
Israel, as well as in many other places, was promoted by the import
and assimilation of foreign, exotic, or global practices into local cultural
spaces, alongside an ideology that advanced consumer culture and a
bourgeois-bohemian global lifestyle.[18] These middle-class groups have
been characterized according to multiple terms and depictions, such as
"the new petite bourgeoisie [that] comes into its own in all the occu-
pations involving presentation and representation," "the post-industrial
middle classes," or a "new middle class that developed around the core

Figure 3.1. "Green Field: Organic Vegetables"—A sign at the entrance of a CSA farm with a depiction of "Zionist pioneers." Photo by the author, December 2019.

of the design, processing, programming and elaborating information and data."[19] Referred to as pro-neoliberal and globally mobile groups, this "new middle class" consists of professionals, managers, technicians, and self-employed "creative" workers "who . . . have been identified with the restructuring of industrial capitalism into its current post-Fordist, global phase as a 'knowledge economy'"[20] and who are preoccupied with "immaterial labor."[21] The notion "creative class" has been widely used as a term that explains the formation of groups affiliated with the emergence of flexible, neoliberal, knowledge-based, and project-oriented capitalism; those who are preoccupied with direct exchanges with clients, employing communicative and affective skills, creating new knowledge, or using "existing knowledge in new ways."[22] It should be noted that, according to urban scholar Richard Florida, in the early 2000s, 20.5% of the workforce in Israel belonged to the creative class, placing Israel 18th out of 25 countries with such a group.[23]

Concomitant to the spread of notions such as the "creative class," other notions such as "neo-bohemians" and "cultural entrepreneurs" as well as colloquial notions such as "fashionistas," "trendoids," and "hip-

sters" emerged and were associated with the same groups.[24] The notion
"hipster," which has recently received sociological research attention,
refers to young people who have adopted an urban lifestyle (and often
participate in the gentrification of former working-class and ethnic
neighborhoods in Western cities) and who are constantly preoccupied
in managing and performing a radical and intensive aesthetic lifestyle.[25]
It was suggested that "the hipster" should be deemed a "neoliberalized
entrepreneurial figure at the forefront of urban cultural production, promoting the art of living well: from micro-brewing to fashion, tattooing
to holistic wellness practices."[26] Thus, although industries such as advertising and architecture, the arts, computer software, film, music, and new
digital media are the exalted occupational archetypes of these groups,
they are also associated with many other urban micro-enterprises that
provide goods and services of style and taste to city-dwellers.[27] While
this cultural, aesthetic, and classed phenomenon of the emergence of the
creative class (with varied descriptions and characteristics attributed to
this ontological and epistemological social category) has been discussed
in the urban context, I point to non-urban spheres as spatial and cultural
domains where models of global aesthetic lifestyles are being shaped. I
suggest that Israeli CSA farmers "translate" ideas and practices associated
with the global creative class to the local, agricultural, and rural context,
and thus shape their global habitus.[28]

　　Those who run CSAs in Israel are mostly young. They are Jewish
men and women in their thirties—mostly of Ashkenazi decent, but a few
are affiliated with the so-called Mizrahi ethnic group (or, more precisely,
with the social category that is described as "the new Mizrahi middle
class"[29]). They all voluntarily chose to set up an organic farm and be
involved in organic farming as their primary vocation. Before they became
farmers, many of them were engaged for a long time in leisure activities
or tourism. They often went on a long trip abroad after the completion
of their military service[30]—a trend that has grown since the 1990s and
is currently considered as part of the "normal" path new middle-class
Israelis take.[31] Many of them wanted to "take a break" from everyday life
in Israel, deciding to engage with remote cultures and live outside the
country. India, Jamaica, Australia, and the United States are some of the
places where they stayed as tourists or as "dwelling-tourists."[32] Upon their
return to Israel, they began to engage in agriculture, first as a lingering
experience of "being abroad" or as a continuation of their experience
with exoticism, the consumption of authenticity, and new cultural ideas
external to the Israeli context or to their cultural background. Some of

them described their new occupation as organic farmers as part of their new lifestyle and as a continuation of their previous experiences and acquaintance with varied postmodern cultural elements.

For example, Shira and Doron are a couple who established and now operate a CSA farm in the Central District of Israel. When I met them, Shira told me,

> When I returned to Israel [from the United States and Central America], I was so into reggae. I met Doron at a reggae party. We started our life-journey together, and wherever we lived, we grew vegetables and had a garden. We traveled to Jamaica and gradually the whole thing with Reggae and Rasta[33] grew in us. We began to feel that we were really living it. We felt how Rasta, the gardening and all of it together became our way of life.

Later, when I asked them questions about the establishment of the farm, about the way they run it, and about the number of customers they serve, Shira complained that my questions were too "technical," choosing instead to preach about ecological notions and ideas, as well as on concepts taken from the Rastafarian culture:

> You have to understand, there's a lot of "pass on this thing," one needs to say the truth and act according to one's own truth . . . and truth is both in words and in the earth. What Doron is able to grow is a real truth that should be shared and learned. It is his way to learn the truth and to share it with others. It's the *I'n'I* [she said in English, referring to the Rastafarian concept "I and I"], that's how one should cultivate the land, just grow vegetables and distribute them. It's our daily effort to reach many people while using minimal resources. In short, it's a way to *spread the word* [she said the last three words in English].

Doron, Shira's partner, joined the conversation and said, "Rasta and organic go together naturally. Because what is Rasta? Rasta is [switching to English] *to go back to Zion,* [switching back to Hebrew] to work the lands of your ancestors. It is the idea of being *self-sufficient* and *self-reliant* [he said the notions "self-sufficient and self-reliance" in English]. Rasta is organic, and organic is Rasta."

Other CSA farmers described the times they spent in spiritual centers in India, experiencing living and working in community gardens in North America, Australia, Europe, and other such activities. Their appreciation of the exotic is symbolically reflected in the types of agricultural crops that they produce. Many of them focus on growing unique fruits and vegetables—the kinds that are trendy outside of Israel and not common in Israeli foodways: parsnips, yard-long beans, daikon, Russian kale, collard greens, and more. These types of vegetables are not common in other places where organic food is marketed in Israel—such as organic supermarkets (see Chapter 4)—and they have become emblematic of CSA boxes. The farmers, for their part, know very well not only how to grow these exotic vegetables but also how to cook them, and they share this knowledge with their customers. Reading the content of hundreds of CSA newsletters reveals that many of the CSA farmers are acquainted with global culinary bon ton as well as familiar with trendy discussions in Israeli and global media about healthy eating, farm-to-table, and the like.

Furthermore, the newsletters that Israeli CSA farmers send to their consumers and/or upload to their websites (often distributed under the English term "newsletter" spelled in Hebrew) reflect their high cultural capital, education, familiarity with various writing styles, and their ability to write creatively. While planning their agricultural activities, they favor crops whose cultivation requires high skills, knowledge, and sophistication. In addition to their inclination toward innovative and adventurous agricultural practices, the CSA farmers present technological abilities and an overall "fluency" in new media and digital culture. Many of them design and operate websites, blogs, and online systems, manage digital communication with their customers, and operate cutting-edge agricultural digital systems on their farms. Many of them put time and effort in engaging with social media, with digital photography and visual processing, as well as with the digital branding and marketing of their farms and produce. It is no wonder that some see themselves as affiliated with Israel's self-perception as a global center of high-tech and entrepreneurship as well as the new Israeli ethos of "start-up nation."[34] Eyal, the son of a veteran organic farmer who now runs the family farm and operates a CSA, said,

> My father was one of the first organic farmers in Israel. He was always creative and talented. He belongs to the generation of Israeli farmers who invented the cherry tomato.[35] But today I manage the farm in a more professional way.

Our farm is organic but don't get it wrong—I manage it as if it is a high-tech company. We are constantly renewing the farm and ourselves. We recently set up a new website. My dad does not know what a computer is; he doesn't know how to go into Facebook and doesn't care about Instagram. There is no way to run a business like ours without technological knowledge. I sit for hours every week—to develop our website, to program and develop [digital] applications that will suit our needs. . . . I did not know anything about it [digital technology, the use of high technology in farming practices and in managing a farm], but I learned. And it helps us. . . . Listen, it's a family business, but a lot of creativity and sophistication is needed here.

Also, many of the CSA farmers I met started working as farmers after working in one of the creative industries. Some of them held key positions in the fields of high technology and communication. Nevertheless, they consider their current occupation—organic farming—to be a new career and, often, to be their "real vocation." In accordance with the cultural logic of "the neoliberal Self"—within which subjectivity and everyday life are blurred and personhood becomes a means of production[36]—their personal background, education, and previous professional experiences are often prominently displayed on their websites. For example, when Gilad, a CSA farmer, described himself, he said, "I used to practice gardening as a hobby, just for fun and for personal consumption, for over seven years. But after several years in high-tech and after my last job as an executive vice president at a big technology company, I decided to change direction and go back to being connected to the land." For Gilad the post-industrial and digital immaterial labor (to use Hardt and Negri's terminology[37]) was replaced by the material-agrarian labor of organic farming. "I had some friends there [in a high-tech company]; all of us left after a few years. One of them was fired, another went abroad . . . relocation, two of my friends work in their start-up, and I became a farmer. . . . This farm is my organic start-up."

## "The Taste of the Past" and the Boutiquization of Organic Food

The "local," namely the local space and the local past, is another common theme in CSA farmers' discourse. Though it seems counterintuitive to hear it from those who possess a global habitus,[38] it is clearly revealed from their frequent use of the expression "the taste of the past" (*hataam shel paam*). This expression, which is commonly used to reflect on the collective yearnings for forgotten tastes and the longing for authenticity,[39] appears repeatedly in leaflets, newspapers, and brochures distributed by Israeli CSA farmers. It is also threaded throughout their websites; it is said by them when they are interviewed for popular newspaper articles and even in many of the interviews I conducted with them.

The portrayal of "the past," according to CSA farmers, dates back to biblical times, as evidenced in many of their written texts, which include quotations and references to depictions of vegetables mentioned in the Bible as well as agricultural anecdotes from biblical stories. The "taste of the past" is also associated with notions related to the history and culture of modern Israel. For example, one CSA newsletter stated, "This week you will find in our baskets an Israeli cucumber [*melaphephon Eretz-Israeli*]. But it is not a regular cucumber; you will feel the difference immediately: small, solid, crisp and sweet." Oded's father [Oded is one of the owners of this farm]—a proud Zionist farmer who already retired—says, "this is a cucumber with 'a taste of the past.' "

A cultural affinity to the local place, and to the historical narrative of the local place, is attained (and even intensified) by another practice prevalent in the Israeli CSA farms: the open farm days. Open farm days take place during the Jewish holidays: Passover, as a spring festival, and Sukkot, as a harvest festival. On these days, many CSA members-consumers visit the farm. "They come from all over the country," as one of the farmers put it, despite the fact that most of them come from the areas in which the farm distributes its produce. One of the CSA farmers described the purpose of these open farm days as a means of "making connections between the consumers and our fields. [It aims] to let them feel what it means to be a farmer, to share with them experiences of authentic farming, just like in the good old days." The open farm days are meaningful days of festivity at the CSA farms. Many consumers come with their whole family and consider this visit to the farm a tourist activity and a way to keep their children busy during the holidays.[40] They enjoy tours around the farms and experience (if only for a few minutes) hoeing and harvesting as well as the heat, humidity, and special aromas inside a greenhouse.

During my participant observations in these events, I often heard how the visitors tried to convince their children, most of who live in urban areas, to be attentive to the explanations given by the farm workers and to express their enthusiasm at the sight of the crops and the flourishing fields. One of the visitors told her son during a tour: "You know, where your grandparents live today, once everything was like this—a lot of orchards and open fields, just like here. When I was a kid, we used to run, play, and eat directly from the trees. Back then, everything was organic."

On one of the open farm days I attended, I noticed that vegetables from the farm's yield were served as refreshments for the visitors. The vegetables were grilled on charcoal and served with organic tahini, organic olive oil, and fresh pita bread made from organic whole wheat flour and baked on a large concave frying pan (known as *saj* or *tava*). While cooking, one of the farm workers shouted: "Hey, everyone, come and taste! This saj is the same as the one at the entrance to Abraham's tent [referring to the biblical figure and to the biblical story of Abraham and the three visitors to his tent]." Such "invented traditional" practices[41] performed and discussed in these open farm days seem compelling to the visitors-consumers of the farm. At the same time, the noticeable discourse of the "taste of the past"—whether it is the taste of the near past ("when I was a kid") or a historical-imagined-invented culinary past ("a saj from Abraham's tent")—appears as a sort of performance of what anthropologist Michael Herzfeld called "structural nostalgia," namely the longing for "the time before time,"[42] in which historical events are "collapsed into generic, imagined, and stylized account of 'the good old days.'"[43]

Practicing consumerism is also evident in these events, as sales fairs often take place in conjunction with the farm days. During one of these events, I saw numerous stands offering a wide range of goods. Next to those stands—right next to the garden—I saw a row of massage beds and a long line of visitors waiting to enjoy treatment for stress or pain relief. Other visitors strolled by stands of a temporary fair set up on the farmland and bought handicrafts (jewelry, ceramics) or second-hand and handmade clothes, enjoyed practicing yoga, attended lectures on "natural education" and "healthy eating," and listened to live music. Thus, the experience of farming and returning "back to the land" were intertwined with a shopping experience. It seems, therefore, that the Israeli CSA farms are compatible with one of the important cultural aspects of modernity and postmodern consumer culture: the search for authenticity.[44] As such, they are shaped—particularly during those open farm days—as hubs within which one can experience uplifting emotional feelings of tasting the authentic past while also consuming pieces of agricultural authenticity.

Israeli CSA farmers describe themselves as significantly distinct from both conventional farmers and organic farmers engaged in large-scale production ("corporate organic"; see Chapter 4). They often mention that they are motivated by a strong urge to follow "their inner [authentic] truth," as one CSA farmer put it. In their perception, large-scale organic farmers, not to mention conventional farmers, are tangled up not only in an unjust or environmentally hazardous agricultural system but also in standardized and mechanical agricultural-industrial production systems (which are "boring and uncreative," as one of them told me).

For these individuals, being a CSA farmer is a calling. It never stemmed, allegedly, from economic constraints, a need to find an occupation, or a way to make a living (at least that is what they said when they introduced themselves to me, when I talked to them, when telling their personal stories, or when giving a talk in a public event or in occasional social interactions, to use Goffmanian terminology[45]). Their farming practices are always discussed as artisanal work[46] and thus as involving not only farming skills but also creativity, deep interest, and an attempt "to express oneself through farming," to use the words of Udi, a CSA farmer from central Israel. Their expertise in producing agricultural produce is tantamount to that of an artist, and their produce has been presented as products of culinary uniqueness and high value. Thus, they seek to be appreciated for not being motivated only by considerations of profit. Following the concept of "art for the sake of art,"[47] they perform "agriculture for the sake of agriculture," or a sort of agriculture in the service of the environment and humanity.

For example, the agricultural activities themselves (sowing, planting, harvesting, plowing, packing, and other activities) are described in the newsletters as sophisticated practices that require creativity and innovation. Practices that are not related directly to farming (such as marketing organic produce purchased from other growers, working with suppliers, advertising, branding, bookkeeping, etc.) are considered marginal and unappreciated activities, although all CSA farmers spend a considerable amount of time dealing with them. The farmers who are most appreciated are those who "know how to work, know how to put their soul in the box," as a CSA farmer from the center of Israel told me when he described one of his colleagues. Farmers who deal too much with branding themselves or who purchase agricultural produce from their colleagues so that they can fill the boxes promised to their consumers are likely to be granted a low status and often mocked by other CSA farmers as "greengrocers" (*yarkanim*).

Many of the CSA farmers seek to achieve a sanctified status[48]—a position based on a firm belief among their consumers that their produce is both distinct and of high quality—and to be recognized as what one might term "boutique farmers." For example, Yoad, a CSA farmer from the Southern District of Israel, told me,

> Some consumers say, "You are a farmer, right? We buy directly from you, so why isn't the price of your produce lower?" And I explain to them: "If you buy boutique bread, for example, straight from the baker, say from the hands of Erez Komarowski [a well-known Israeli chef who is famous for baking boutique bread], you will pay more than the bread you buy in the supermarket, right? If Eyal Shani [an Israeli celebrity chef] serves a pita with kebab in his restaurants—you will pay more than in Shipudei Hatikva [a casual kebab stand in Tel Aviv], right? The same is true of me. You can buy organic food in other places, in the organic supermarket, for example. It would be organic, for sure, but it's industrial organic and you can't argue with the fact that with large quantities the quality is reduced." I try to explain it to them [to the customers] that what they get from me is boutique quality. Just as there is a boutique baker, there is also a boutique farmer.

A similar perspective that sees the work of the CSA farmers as artisanal work was offered by Meital, a farmer and co-founder of a CSA in south-central Israel:

> There is something about our organic farming, which is just like art, and not everyone knows how to do it. Cultivating cherry tomatoes that are full of taste and sweetness and stay firm, believe me—it is a challenge. And growing an organic cherry tomato—is a work of art. In all modesty, I think that Roi [her partner, the farm's leading farmer] is a professional, an artist who knows how to cultivate such tomatoes and to bring out these good products.

Another expression of the boutiquization of organic food can be found in two of the main farming practices widespread in many organic farms, including in the Israeli CSA farms: crop rotation and crop diversity. These practices are perceived as increasing soil fertility and crop yield,

Figure 3.2. CSA in Israel: "Premium tomatoes—organic." Photo by the author, June 2011.

contributing to the farm's biodiversity, and are credited with ecological and agricultural advantages. However, it seems that these techniques have not only agrarian values but cultural and symbolic values as well. Interestingly, they fit with the postmodern habitus of cultural omnivorousness that many of the Israeli CSA farmers—and their consumers—hold.[49] In this respect, the diversity of crops and organic produce is compatible with new patterns of consumption based on cultural openness and heterogeneity in culinary tastes. Culinary diversity is exemplified in the content of the CSA boxes sent to consumers: Vegetables that are perceived as local and traditional, such as baladi tomatoes (heirloom tomatoes), Egyptian broad bean pods, local black-eyed peas, and local cultivars of *Cucurbita* (squash) are packed alongside crops with global or exotic charm, such as American kale, Japanese spinach leaves, Thai beans, Chinese cabbage (*bok choy*), and Chinese greens (*tatsoi*). The latter were sent to the consumers while emphasizing their (alleged) geographical origin or use in exotic cuisine. In addition, Cinderella pumpkin, spaghetti squash, and many other foodstuffs are described as exotic supplements to "your average salad of cucumber and tomato," to quote one of the CSA's newsletters. Because of the quality and variety of the vegetables, CSA farmers sell some of their produce to chefs who use them in gourmet restaurants in Tel Aviv and Jerusalem. These chefs reward the CSA farmers not only monetarily but also by projecting culinary prestige upon the CSA farms. The farmers, for their part, tend to capitalize on their connection to the chefs and proudly discuss these connections in their newsletters, websites, and conversations with consumers.

Figure 3.3. Boutiquization of organic food: Small-scale organic farm, Sharon Plain. Photo by the author, November 2019.

Figure 3.4. Boutique farming: Small-scale organic farm, Sharon Plain. Photo by the author, November 2019.

Many of the CSA farmers themselves have extensive culinary knowledge, and some of them even have professional cooking experience. For example, Meital, co-founder of a CSA in south-central Israel, was trained at the Cordon Bleu Hotel Management and Cooking School in Paris, and Liron, a CSA and organic farm owner who branded himself as the "organic chef," graduated from a training program in culinary arts in the United States and worked for several years in gourmet restaurants before he moved back to Israel and became an organic farmer. Equipped with vast culinary knowledge, CSA farmers enrich their newsletters and boxes with recipes for sophisticated and fashionable dishes. Thus, the value of the vegetable box, which at first glance may look like a collection of vegetables, increases immediately with the exclusive recipes that come with it: "Okra and Summer Vegetables in Coconut Milk," "French Potato Gratin," "Young Carrot Cream Soup," "Caramelized Stuffed Onion with Organic Honey," "Gazpacho Soup," "Spinach and Citrus Salad," "Butternut Squash and Purslane Steamed in Butter," "Pumpkin Pie with Candied Pecans," and many other recipes (vegetarian or vegan), which are all based on the vegetables from the CSA farms.

No less interesting than the content of the newsletters is what is not included in them. A cumulative reading and a content analysis of the newsletters collected from various CSA farms reveal that texts that deal explicitly with ethical, social, or environmental aspects embodied in organic food or organic farming are rare. The same is true for radical or critical terminology, encouragement of social or environmental activism, or even references to general or burning political issues. For example, during a time period when there was a heated discussion about the credibility, meaning, and quality of organic agriculture in Israeli mass media, none of the CSA newsletters I examined mentioned the issue. Similarly, when issues related to social justice, environmental care, and political economy in Israel were intensively discussed in the public sphere and could be connected to the discussion about the meanings of organic and the importance and benefits of the CSA model (for example during the middle-class protests in the summer of 2011), these issues were never mentioned in the newsletters. When I asked about this, trying to understand why many of them refrain from expressing an explicit political stance—and thus depoliticizing organic agriculture—many sought to separate their political positions or civic engagements from their agricultural and commercial activities. For example, Danit, a farmer who operates a CSA farm in the central section of the Coastal Plain of Israel, explained,

> I don't want to "go against." I don't want to be negative in the newsletter. . . . I don't like to hold demonstrations or riots, that's not what I do. I also don't like dealing with clichés, talking about romantic ecology, and all that stuff. These things can be heard elsewhere. Don't misunderstand—from a personal point of view—it does interest me. It bothers me very much what is happening here in the country and the brutal capitalism and all that. But I will not write about it in the newsletter. . . . I don't want to talk too much to people about theories, why organic is important and all that. I'm much more interested in writing about the family, about the day-to-day, about our special carrots . . . about our sweet Thai beans that started growing, about what we ate, what we cooked. These kinds of things. This is also what interests our customers.

It should be noted, however, that some of the critical political principles on which the CSA model is based are explicitly expressed in the Israeli version: CSA farmers contribute to promoting the discourse about local farming and the need to reduce the use of synthetic substances. They provide transparency about their processes of growing, packaging, and marketing, and even work hard to strengthen social embeddedness[50] by establishing social relations with their consumers and trying to get to know them personally. But all these critical discursive elements are delivered within the cultural frames of artisanal ruralism and as symbols of luxury and cosmopolitan taste. In this framework, organic food is shaped according to global and fashionable similes of "community" and "locality" as well as local history and tradition. The latter serve as important components, aspects, and values of the "goodness" of the organic food produced in CSAs (in relation to other organic foods, namely those produced and marketed in other venues). In doing so, the Israeli CSA farmers not only foster the commodification and boutiquization of organic food but also create an inextricable interlacing between organic food and cosmopolitanism, global habitus, status, and refined taste. Some of these meanings also arise from the analysis of the discourse and praxis of Israeli CSA consumers as described below.

## Israeli CSA Consumers

Many Israeli CSA consumers are familiar with the term *community-supported agriculture*. Nevertheless, they rarely identify themselves as "community members" or as "agricultural supporters." For many of them, the engagement with CSAs is only one option for organic produce consumption. "We're just getting an organic box directly from the farmer to our home," some CSA consumers claimed in my interviews after having asked them what it meant for them to be CSA members. These meanings might stem from the ways in which CSA operates in Israel. According to the Israeli model, customers simply subscribe to the farm, receive a weekly organic vegetable box, and pay a monthly fee for the vegetables they consume. They do not contribute to the establishment of the farm, do not share in its management, and very few of them take any active part in the growing process. Furthermore, in most Israeli CSA farms, the interactions between growers and customers are mostly done through the CSA's website and through the weekly newsletter distributed by the farmers. Thus, it seems that a sense of community is created not through a physical meeting but rather through the mediation of digital-textual information.

Israeli CSA consumers are not referred to as "subscribers" or "members" by the growers, or even as consumers. Rather, they are referred to as "families." Most of them are middle-class professionals, residents of major cities or suburbs of the major cities or settlements in the wealthiest municipalities in Israel (including the wealthiest settlements beyond the Green Line), Ashkenazim, or those who might be considered as part of the new Mizrahi middle class.[51] And indeed, most of them subscribe to CSAs as families. For example, when I looked at the consumer list of some of the CSA farmers, I saw that the consumers are written as families and not as individuals: the Cohen family, the Friedman family, the Katz family, and so on. Several scholars have already pointed out that family meals often serve as a means of structuring the social categories of family and home.[52] In Israel—similar to other Western and Westernized societies—family meals are increasingly abandoned due to the growing habit of "eating alone" and the increased frequency of "eating out."[53] This process leads to a decrease in parental control, a sense of insecurity, and lack of knowledge about the quality of the food that children eat. These feelings are in line with the prevalence of public dialogue on risk in contemporary civic life, as well as with sociologist Ulrich Beck's understanding of the emergence of a "risk society," in late modernity.[54] Therefore, it seems that consuming vegetables from a CSA represents "safe consumption" in a threatening and unreliable food system. Anat, a 42-year-old mother of three children, described this feeling:

> Do you know why I insist on [buying] organic [food] and why I order from Yossi [a CSA farmer]? Because I've read studies that show that by age five, all the development in a child's brain takes place, so I decided that until the kids are five—I'm guarding them. I nourish! I'm a warrior! I'm a lioness! I will do all I can do so that they get the best nutrition possible. . . . I trust Yossi. I've visited his farm and I've seen how they work. I don't trust the large marketing [conventional] chains, and I don't trust Eden [i.e. Eden Teva Market; see Chapter 4] either.[55]

Several consumers stated that their preference for consuming organic food via CSA is derived from the desire to support local and small-scale individual farmers. In the words of Oren, a 42-year-old male CSA consumer from Jerusalem, "The problem with conventional [agriculture] is a global problem, and not only in Israel. So, I buy organic food from Gilad [the name of a CSA farmer]. It is not cheap, but I do it, first, for my kids. Second, I like to think that I support Gilad. I [like the fact that] I know where my money is going." Similarly, Hila, a 40-year-old female CSA consumer, itemized the advantages of buying directly from a farmer as "comfort, quality, and taste." Then she added, "and by the way, my money goes to people I trust." Therefore, although Oren's and Hila's words indicate a certain interest in supporting private growers, they emphasize health and trust as their main motivations for being CSA consumers.

Some of the CSA consumers mentioned that they prefer to buy "local." However, those who emphasized the values of locality were the ones who are well traveled. For them, the connection between "organic" and "local" was formulated while traveling abroad or living outside of Israel for prolonged periods (e.g., relocated high-tech workers, academics, and businesspersons). Yael, a 40-year-old female CSA consumer from Tel Aviv, said (in Hebrew spiced with American English):

> When my baby was born, we lived in New York. So, I decided she would not get anywhere near all the junk food they eat there. We heard from friends and looked for *community-supported agriculture* [she said in English]. We found it easily! It was organic, *like real*! [she said "like real" in English]. It was local, communal—a real CSA. We would go to visit the farm on [turning to English] *harvest festivals*. When we moved to Sacramento [switching back to Hebrew], we found a CSA. There was no problem finding, there were a lot of them. Our [CSA] had a huge, amazing vegetable garden.

Thus, paradoxically, Yael and other CSA consumers became acquainted with consuming "local" food not while they were within Israel but while they were abroad. Their sense of locality and being local, then, is related not to rootedness nor to connectedness but to the experience of being global subjects.[56] Yet another example is Rotem, a 38-year-old female CSA consumer from Jerusalem, who described how organic, local, and exotic vegetables remind her of being abroad:

> The CSA in California was especially cool. It was fun cooking there! What vegetables! Fresh and local heirloom tomatoes . . . we got all the bizarre vegetables that grew there. Who even knew what heirloom tomatoes were?[57] It really challenged

me, culinarily speaking. Or this vegetable, "kale." We got it
here [in Israel] many years later, from Gilad [the CSA farmer
from whom she buys]. I remember—what an excitement it
was! . . . It really reminded me of California.

The vegetable known as "kale" indeed grows on Gilad's CSA farm near
Jerusalem and is therefore considered "local" in the geographic sense,
but as evidenced in Rotem's words, its "locality" is recruited to confirm
her affiliation to and longing for California as well as her cosmopolitan
identity. In this way, the various political (civic) meanings embodied in
the "locality" of organic food (environmentalism, anti-corporate ideas,
community, etc.) are muted and replaced by a global (consumer) discourse of "quality," "freshness," and "distinctiveness" (the uniqueness of
the kale, for instance).[58]
    The discourse of CSA consumers can also be seen as an expression
of the "boutiquization" of organic. Yehuda, a 39-year-old CSA consumer
who works in high-tech, said,

> I travel a lot overseas on short business trips. It is from there
> I've known of the organic vegetable scheme for a while. [Here,
> in Israel] I get it [a weekly organic vegetable box from a
> CSA] because it is so much fresher. . . . As a kid, I grew in
> a moshav [cooperative settlement]. For years I've longed for
> vegetables this fresh. When I get home from work and the
> box is waiting by the door . . . I pick up the box and like I do
> with a good glass of wine, I shake and smell . . . ah . . . the
> smell. . . . It reminds me of my childhood when we would
> eat [vegetables] straight from the field.

Thus, although to some extent Israeli CSA consumers are interested in
civic values (such as supporting local private growers rather than large
corporations), it seems that these values are mostly secondary to personal
concerns when they choose to consume organic food from CSA farms.
    This symbolic interlacing of organic, cosmopolitanism, and good
taste is realized not only by relations between rural-peripheral actors
(CSA farmers) and the Israeli bourgeois-suburban families who consume
their produce but also by the relations between urban actors and urban
consumers, such as those who are found in Israeli farmers markets—which
will be discussed below.

## "Every Tomato Goes Through an Audition": Israeli Organic Farmers Markets

Much more than merely sites for shopping, food markets—especially those that operate in urban spaces—are public spaces where vendors, shoppers, and the occasional visitor develop social relations and cultural affinities.[59] Sociologist Liora Gvion contends that ascribing notions of freshness, naturalness, and authenticity to urban food markets turns them into enchanted spaces where "stall owners' skills and customers' desires merge to make products appear more pleasing to the eye."[60] Historically, many open-air markets and bazaars operated in the territory known as Israel/Palestine. Following the Jewish immigration to Palestine, markets began to operate mainly in urban places. Among them are the open-air markets that were established in the 1920s in the city of Tel Aviv: Levinsky Market and Carmel Market. During the 1960s, when retail markets shifted to supermarket chains (see Chapter 4), and later, in the 1980s with the rise of the shopping malls and the power centers (unenclosed shopping centers),[61] the status of urban markets declined. In recent years, however, the experience of shopping in an open-air market has again become trendy. The Carmel Market in Tel Aviv, the Mahane Yehuda Market in Jerusalem, the Ramla Market, the Netanya Market, and others are places that have been "renewed" and gentrified. Nowadays, those markets—which were once considered common and accessible to people from all walks of life—have turned into "culinary playgrounds" for members of the middle class. These middle-class consumers, who are often the beneficiaries of the global economy, appreciate the feeling of shopping and being in a space that, allegedly, is not yet colonized by corporate values.[62]

Concomitantly—following of the increase of criticism against industrial agriculture as ecologically, socially, and economically destructive, as well as the advocacy for creating sustainable and just alternatives—an interest in farmers markets rose in many places in the Global North.[63] Through shopping for locally grown produce in urban farmers markets, consumers are often encouraged to contribute to what is believed to be green economy, sustain small businesses, and practice meaningful social relationships.[64] Organic food has become an integral and important component of farmers markets in Europe and North America. Considering the vast social meanings attributed to farmers markets in global mass media and in public discourse, one can assume that they realistically promote the re-embeddedness[65] of social relations and function as sites that hold off processes of neo-liberalization and globalization. However, critical studies claim that farmers markets, and their socio-environmental economic strategy, operate according to the exact logic of neoliberalism. These studies point to the contradictions between the farmers markets' normative aims and the individual-oriented economic strategies used by them and to the fact that social and environmental change opportunities are, essentially, fostered in these places through market behavior.[66]

The interest in and popularity of farmers markets have also revived in Israel and have further expanded following the success of the Tel Aviv Port Market (later transformed into The Port Market, known in Hebrew as Shuk Hanamal). The market was founded in 2008 by three young entrepreneurs and foodies: Shir Halperin, a French-trained chef and food journalist; Roee Hemed, an architect and entrepreneur; and Michal Ansky, who graduated from the University of Gastronomic Sciences in Northern Italy (an international academic institution established by Carlo Petrini, the founder of the Slow Food Movement). Ansky is currently a famous Israeli food celebrity after being a judge on the Israeli version of the reality TV show *MasterChef*.[67] She is also the daughter of Sherry Ansky, a famous cookbook author and food journalist. As a direct continuation and implementation of Ansky's studies in Italy, she strives to "bring Israeli farmers back to markets," to use her own words.

The market the three young entrepreneurs initiated was supported by the international Slow Food Movement. Following its success, their farmers market project expanded and multiplied into branches in other cities such as Be'er Sheva, Ra'anana, Herzeliya, Holon, and Rishon LeZion. And just like farmers markets in the Global North, organic food is an important element in the culinary repertoire of these Slow-Food-inspired markets. This is best illustrated in the "vision of the market" as it appears on the Tel Aviv Port Market website:

> Inspired by food markets in Europe such as La Boqueria in Barcelona, the Tel Aviv Port Market takes its food cues from the Slow Food Movement which prioritizes and prefers clean and fair produce. [The Tel Aviv Port Market is committed to] promoting eco-responsible farming of plants, seeds and livestock, emphasizing organic, seasonal and local produce. . . . Green agenda, environmental ideology, atmosphere, ambience and quality are [our] pillars and our key words.[68]

"Organic" and "environmental ideology" are notions that were also mentioned in an interview I conducted with Ansky. When I asked her what were her reasons for including organic food in her farmers market, she answered,

> Basically, it is an ideological act. . . . The idea is to strengthen the local economy. . . . Real organic food, such as the foods that one gets to eat right from the hands of the farmer—this is the food that I am talking about, that is my ideology. And I'm not talking about organic food that one can find in the organic supermarkets, where you buy vegetables that are almost rotten, those that were harvested two weeks before one sees them on the shelf. It's not only about not being sprayed [with pesticides]. I'm talking about delicious organic! Fresh! Local! Seasonal! . . . Buying organic food should also be a cultural experience, a shopping experience.

It seems that much more than promoting ideology, the atmosphere and ambience of the market is that of a "shopping experience"—more precisely, a cosmopolitan shopping experience. For example, *Al Hashulchan* magazine (literally, *On the Table*), an Israeli magazine devoted to food, wine, and culinary culture, chose to describe the market in this way:

> This is the market we dreamed about when we travelled to Barcelona, Rome, and London, and asked ourselves: "Why don't we have one like this [in Israel]?" The farmers market in Tel Aviv Port, which is expanding and now operates in five other cities, expresses something new and different—it is cosmopolitan and glamorous but also close to nature, to agriculture, and to the seasons of the year.[69]

When visiting the market, one hears pop music immediately upon entering the compound where it is located. A colorful stall of juices is right at the entrance, and the first thing many visitors do is buy a bottle of natural fruit juice. Young peddlers working in food stalls smile at the visitors. They all dress in chef's uniforms and wear fashionable kerchiefs (bandanas). "The inspiration of food markets in Europe," as reads in the above-mentioned "vision of the market," is easily noticeable: next to the entrance one can find a cheese shop that offers "cheeses and products from all over the world alongside local produce" (as is written in chalk on a board in front of the store). In a nearby delicatessen, one can see a prominent inscription that says, "An Italian delicatessen that brings aromas and flavors to the Israeli households directly from Italy and Spain," as well as imported wines, fresh oysters imported from Western France, dry pasta from Italy, and many other products—all of which are non-local but nevertheless found their way into the "local" farmers market.

Alongside the stalls with the imported foods and next to other stalls that offer high-quality (and high-priced) local conventional fruits and vegetables, one can see a large store that exclusively sells organic food. The vegetables and fruits in this organic store are neatly arranged in crates on shelves that hang all along the walls of the store. This design seems to be homage to the old open-air markets. Another possible interpretation to the design of the store is the creation of a simulacrum—*a representation* of being in an open-air market and a vision of "a market from the past," to use the terminology of Jean Baudrillard.[70] Interestingly, this design serves as a simile in the space it is supposed to signify: the crates that signify an experience of being in an authentic market are within an actual market, the farmers market in Tel Aviv Port.

Nir, the owner of the store and the one who is responsible for its design, told me,

> I searched for my Self for many years. I studied, traveled abroad, worked in sales departments in high-tech companies, but after this long journey searching for my Self, I turned to grow and sell vegetables, because I am the son of a farmer. Much of the produce you see here I grow by myself in the fields of the coastal plain south of Rishon LeZion.

On the store's website, he wrote, "Previously only a few people knew about the qualities of organic agriculture, but it is now the order of the day. Buy organic—that's the right thing to do." However, it seems that the organic food in his store is not only there because it is "the order of the day." Instead, it is mostly "the fashion of the day." Nir's discourse and practices seem to tie organic food with values of high quality, luxury, and aesthetic culinary trends. He presented himself, and his store, as "different from the vendors and the other stalls in the market," saying that his organic produce is "real, authentic and healthy"—terms he used repeatedly during our interview. However, the fashionable and updated design of his store serves as a strategy for its integration into a fashionable and popular farmers market in Tel Aviv Port. Indeed, his produce—just like all produce offered for sale at this farmers market—looks extraordinary. There is no doubt that the fruits and vegetables are carefully selected—all fresh, firm, shiny, and arranged in an eye-catching way. "Here things must be very fresh and beautiful," he says. "Every tomato goes through an audition before it reaches the shelf."

The high price of his produce, the touristic location, and the similes of gourmet culture all imbue the organic food with elitist distinctive meanings. "When people buy in my shop, they don't look for low prices. My consumers are of the high class, they are those who ask themselves: 'What is the best vegetable I can get? Organic? So, I buy organic.'" Similarly, Amit, the owner of an organic stall in a branch of the farmers market that operates in Ra'anana (a city in the Sharon Plain of the Central District of Israel) provided an intuitive Veblenian analysis[71] when I asked him about his customers: "These are the people who want to tell their friends and family, 'Look, I cooked a whole meal with organic food.'" He also told me that he always makes sure to have a certain amount of unique and exotic vegetables available, some of them at prices that are not for everyone. "Some people like new and special things. And they can find it here, at my stall. These specialty vegetables [he said this while pointing to a bunch of rainbow Swiss chard] are a magnet to the people who visit the market looking for new things."

It should be noted that organic traders such as Nir and Amit do not underestimate ecological issues: they insist on certified organic produce and perform environmental care by providing consumers with reusable baskets made of woven straw for use while shopping at their stores and stalls, by encouraging the reduction of the use of plastic bags, and by raising environmental issues in conversations with their customers. For example, during one of my observations in the farmers market in Tel Aviv Port I saw a customer who used too many plastic bags and was reprimanded by Nir (in a kindly, but nevertheless determined, manner): "Come on, why, why are you doing this, isn't it a shame? Don't you care about my expenses and about our environment? Don't you care about our globe? Please, use the basket instead of these plastic bags!" Nevertheless, most of the conversations with the consumers deal with the uniqueness of the produce, its quality, and its freshness. "My customers are not devout followers of organic," said Nir. "It's just a bonus to them. They have money and they know what organic is, they believe it's good for them, so they'll just buy it. But they will never compromise on the quality."

The emergence of these stylish farmers markets, and the manner in which organic food is marketed in them, indicates the assimilation of a global foodie culture in the Israeli culinary sphere and among affluent groups in Israeli society. The foodie culture, as described by sociologists Josée Johnston and Shyon Baumann,[72] is composed of three main trends: (1) the rise in the popularity of ethnic cuisines; (2) the search for specialty commodities and premium ingredients;[73] and (3) the spread of local, organic, and sustainable foods. In this regard, Nir and his colleagues, the organic traders at the Israeli Slow Food–inspired farmers markets, provide their customers with a venue that serves to practice at least two of these trends through consuming organic and specialty foods. "I offer the Mercedes of the vegetables," said Nir. "They are willing to spend money. You see, buying organic food in a farmers market is a status symbol, just like driving a Volvo or Mercedes. I can't deny it, this is what is going on here with my customers, there is a lot of façade here. They come here as if they were in Paris or San Francisco, not near Jaffa."

## "Orbanic" Consumers

Another farmers market in Tel Aviv that provided its visitors with a feeling of luxury and "being abroad," one in which only organic produce was offered for sale, was the Orbanic Market.

Since its inception, this market was infused with cosmopolitan images. "[Organic farmers markets] thrive in Amsterdam, Berlin and Paris, and now . . . also in Israel!" was written on flyers distributed weeks before the opening of the market. The name of the market, "Orbanic"—a portmanteau of the English words "organic" and "urban"—symbolizes the founders' intention of inscribing a sense of modern global urbanism onto the space. The market was established in the area where the Menashiya neighborhood in Tel Aviv was located, in a place known today as Hatachana (The Station). In 2010, this place—where the old train station that connected Jaffa to Jerusalem operated (1892–1948)—went through a process of gentrification and became a compound for recreation, leisure, and shopping and a part of a broader central business district. The Orbanic Market operated there between 2010 and 2011. It included 40 decorated and beautifully arranged stalls. Each stall prominently displayed a certificate from an organic supervision company attesting to the credentials of the food products. In addition to buying organic foods, the visitors to the market were invited to join tai chi practice or attend one of the regularly hosted workshops that discussed issues such as personal health, nutrition, or environmental care. "World Music" (music incorporating elements of traditional music from the developing world) was heard from large speakers set in the center of the market avenue and created a relaxed ambience.

When I first visited the market, its visitors seemed to me to be enjoying the atmosphere it offered. Here, for example, is a description of one consumer's impressions as she described on her personal blog:

> Friday morning [and] I had a tight schedule: first—a visit to the organic market, then a yoga class. [At the Orbanic Market] I met a friendly and smiling stall owner. We start an organic conversation and some tastings of a special kind of purslane [edible weed] that comes from France. . . . It's been a long time since I've seen such crisp purslane. Purslane leaves are healthy and contain a lot of omega 3. Later, I was introduced to "spaghetti squash"—a pumpkin rich in carotenoids, which can protect you from heart disease, diabetes, and even cancer. I added it to my ecological basket [reusable grocery bag]. On another visit [to Orbanic Market], my eyes were attracted to the purple eggplants. I imagined how great they would be in my Sicilian pasta. I also bought zucchini and leeks that go well in the same dish[74].

This description represents the associative and symbolic range from which the "organic discourse" in Orbanic Market was composed. It points to the health and culinary virtues of the vegetables ("purslane leaves are healthy and contain a lot of omega 3," "protect you from heart disease, diabetes and even cancer"), emphasizes their exotic uniqueness ("a special kind of purslane that comes from France"), and highlights the global gastronomic experience they represent (Sicilian pasta), and the use of a fashionable cloth bag is presented as an act of ecological significance.

Geographers Benjamin Coles and Philip Crang maintain that naming a space in which food products are sold as "local" gives the consumer a sense of doing something ethical. This, according to them, creates both a de-fetishization and a re-fetishization of food. By revealing several aspects of the production process—for example, identifying farmers or places of origin—it de-fetishizes. Yet, by hiding other exploitative social relations embedded in the production process, it re-fetishizes.[75] Accordingly, the social relations in Orbanic Market promoted a de-fetishization of organic foods, mainly through the social encounters between consumers and growers. At the same time, the coupling of the notion "organic" with the notion "local" in this market expressed symbolic re-fetishization. While wandering in the market, I noticed that consumers and vendors alike responded to the imperative that encouraged them to engage in "locality." Consumers often asked about the source of the crops, and the vendors responded willingly and spoke about their lives in the rural areas. Thus, during conversation, organic foods, both the Israeli produce and the imported organic produce, were loaded with meanings of localism. However, paradoxically, these communicative interactions between organic farmers and consumers—which are perceived as an engagement in "locality"—strengthened the global symbolism of the market itself. Here, for example, is a text written in an online magazine that deals with topics of lifestyle: "Organic Market in 'Hatachana'—Buying Local: Equipped with my shopping cart I go shopping, like in a small village in Italy or Greece: fresh produce in small stalls."[76] This description illustrates how the consumption experience in the market was perceived: hedonistic, trendy, young, vibrant, and touristic.

The emergence of Orbanic Market in Tel Aviv demonstrates the ways in which organic food is integrated in the creation of a new-cool-culinary-urban atmosphere encapsulated in cities like Tel Aviv, Jerusalem, and Haifa. This trend is reflected in a practice prevalent among young urban residents known as "doing market" (*lahsot shuk*). The original meaning of *lahsot shuk* referred to the practice of shopping on the eve of the Sabbath by the middle and lower classes. Nowadays, this notion is associated with the new fashionable culinary trend, which includes practices of eating gourmet food in the market and the emergence of gourmet restaurants that operate in the large urban markets. Along these lines, and probably as part of this trend of "doing market," many of the visitors to Orbanic Market went there for recreational purposes and not necessarily because of its "organicness."

Gradually, however, fewer and fewer consumers visited the market until it stopped operating. Some consumers argue that the reason the market closed was because of the atmosphere, which seemed to them as "fake, too clean, and inauthentic," as one told me. Thus, the only farmers market in Israel that was based on a direct connection between organic farmers and consumers, the one that was based exclusively on local and certified organic produce, was perceived, eventually, as inauthentic and an imitation of a "real" market.

The symbolic connection between culinary culture, authenticity, and "organicness" was actually accomplished in the farmers market in Tel Aviv Port and in the newly fashionable open-air urban markets where various foods are available. Among these markets is the renewed Carmel Market in Tel Aviv or Mahane Yehuda Market in Jerusalem. In Carmel Market, for example, one can find vegetables, fruits, eggs, cheese, spices, legumes, and baked goods. There are several delicatessens that offer pickled fish, non-kosher meat, olives, pickles, imported cheese and canned goods, and merchandise from either the Far East or the former Soviet Union. Additionally, one can find espresso bars that sell expensive brands of coffee beans, and on the road parallel to Carmel Street, there are meat shops and fishmongers as well as bistros and eateries—all of which offer seasonal menus and claim to use only produce from the market. In addition to all this, two branches of organic food stores can be found on the main street of the market. In these renewed open-air markets, organic food transforms into "boutique food" alongside the other non-organic gourmet and exotic foods.

One of my observations took place in an organic store-stall that opened in the Carmel Market in Tel Aviv. The store is located on the main street of the market across from a new, lively café and a shop selling local and imported cheeses. In the store, I met a few consumers I knew. I recognized them and remembered that they had frequented the Orbanic Market before it closed. Most of them said they didn't really care that the Orbanic Market had stopped running. They actually noted that they preferred to do their grocery shopping at this store, in the Carmel Market, where they can buy organic food, as well as "unique" and "exotic" foods, and enjoy the "authentic atmosphere of the Carmel Market." For many of these urban organic consumers, the notion "organic" plays only a secondary role in their culinary cultural repertoire and in their local interpretation and adaptation of the global foodie culture.[77]

For example, Ido and Meital, two young Tel Aviv residents (both around 25 years old), told me,

> Every Friday we ride our bikes to the Carmel Market. There, we have breakfast, buy organic vegetables and some special pasta. . . . We host our friends for dinner very often. We prefer healthy foods. Sometimes we go to the farmers market in Tel Aviv Port. We don't go into Eden [Teva Market]. It's such an elitist supermarket! We also don't order a delivery of organic boxes [from CSA]. We prefer to buy here in the [urban] market. It's much more authentic and much more fun.

Their words illustrate the integration of organic food in the culinary repertoire of urban upper-middle-class groups in Israel, those who are characterized by high cultural capital and who distinguish themselves from mainstream culture. These consumers belong to the social categories that have been dubbed the "leisure class," "BoBos" (bourgeois-bohemian), "creative class," or "pseudo-anti-capitalist hipsters"[78] as discussed earlier. Sociologist Sharon Zukin describes the renewing of urban food markets as a process of constructing "oases of authenticity," a process done as part of a broader gentrification and construction of affinity between these social categories and urban spaces.[79] Therefore, the organic foods offered for sale in the Israeli open-air urban food markets and conceived as authentic might contribute to the creation of a hub of authenticity in Tel Aviv. In contrast, farmers markets where exclusively organic foods were sold without any additional "authentic" or "gourmet" foods (such as Orbanic Market) attract only a small number of consumers.

Taking everything into account, organic food—as it is produced, marketed, and consumed in Israeli CSA farms and farmers markets—serves as a means of demonstrating cultural openness to the world, creativity, technological skills, digital literacy, and neoliberal communicative and affective skills. However, it also serves as a means of expressing what seems to be the opposite: a longing for local and traditional-cultural aspects. These aspects contribute, all together, to the construction of a global habitus and cosmopolitan identity among suburban and urban groups of the upper middle class in Israel. The following chapter, which focuses on distribution processes, reveals that the locus of organic food is not only urban and suburban areas but also kibbutzim and peripheral spiritual communities.

Chapter 4

# Organic in the City,
# Organic in the Kibbutz

## Spheres of Distribution

In his *Grundrisse: Foundations of the Critique of Political Economy*,[1] Karl Marx explains that distribution activities in capitalist markets, and in any given commodity system, are by no means simple acts of the circulation of products: "Distribution steps between the producers and the products, hence between production and consumption, to determine in accordance with social laws what the producer's share will be in the world of products."[2] Distribution, according to Marx, "is itself a moment of production . . . in that the specific kind of participation in production determines the specific forms of distribution."[3] In accordance with the Marxist perspective, political economy scholars Jonathan Nitzan and Shimshon Bichler point out that corporations operating in contemporary political economic systems—including retailers and distribution companies—work constantly not only to provide goods and services but also to shape and reshape markets, politics, and culture.[4] Spheres of distribution in any given food system or culinary field—shops, supermarkets, restaurants, food packing and delivery firms, and the like—are important places and institutions with the closest link to consumers (compared to farmers or food manufacturers). Powerful participants in these spheres influence which commodities are perceived to be acceptable and which are available as well as construct and communicate the symbolic and exchange (economic) values of crops and food products, codify nutritional and cultural meanings of foods, and negotiate consumers' tastes.[5]

101

Distribution and marketing activities in fields of organic food (including the field at issue here) are no different. While the previous chapters describe the emergence of organic farming in Israel (Chapters 1 and 2) and discuss the structural and cultural position of farmers and producers in the contemporary Israeli field of organic food (Chapter 3), this chapter focuses on the spheres of organic food distribution, on the ways in which marketing methods take place, and on the ways in which organic food purveyors, distributors, and retailers shape and reshape the field of organic food in Israel.

Farmers markets and box schemes (CSAs)—such as those discussed in the previous chapter—are often seen as the quintessential sites of organic food distribution, allowing consumers to purchase organic, seasonal, and local foods. Together with farm-to-table restaurants, family farm shops, food coops, and small grocery stores, they are considered the ideal-types of organic food distribution forms. Additionally, they are highly regarded as sites in which ideas related to the alternative food movement have been molded.[6] However, the majority of organic sales in the Global North and Western countries take place at supermarkets or in big-box stores, and not through such small-scale distribution initiatives.[7] In tandem with the conventionalization of organic farming in North America, the distribution structure of organic food was corporatized and alternative agricultural initiatives were absorbed (both materially and symbolically) into corporate institutions. Consequently, a large portion of organic foods that originate in large-scale industrial farms are packaged and transported by third-party suppliers and distributed within global commodity chains.[8]

A vast amount of literature has documented the impact of the significant change organic food distribution has undergone. This transformation—from a small-scale distribution system to a large-scale distribution system—has raised questions about the commitment of social engagement and ecological responsibility embedded in these new methods. "Corporate organic"[9] systems in the Global North have provoked particularly harsh criticism regarding the impact of their distribution structure, which is based on the transport of organic food within global food systems and commodity chains. It is argued that large natural food retailers transport organic food great distances from where it is grown and processed to where it is sold. Considering the fact that conventional supermarkets offer increasingly more organic items, several critical scholars claim that the corporatization of organic has co-opted the messages and ideas of what is conceived as the original organic movement.[10] Sociologist Josée Johnston and her colleagues, as well as other commentators, employ a dialectical approach that recognizes a dynamic relationship between market actors and social movements, and see "corporate organics" as "a hybrid entity that cultivates a fetishized image of ecological embeddedness in locally scaled places, while obscuring long-distance commodity chains, globalized trade, and centralized corporate control over the food system."[11] Accordingly, when referring to the cultural meanings of urban sites of organic food distribution, sociologist Sharon Zukin points to the common denominators of both retail outlets for small family farms (such as the Greenmarket in New York City) and the stores of big retail chains devoted to natural and organic foods. She describes both categories as "spaces of alternative consumption that offer consumers places to 'perform difference' from mainstream norms (as well as from consumption habits associated with working-class and ethnic minority groups), and thus acquire cultural capital and social distinction."[12] Taken as a whole, the academic discussion on organic food production and retail evolved around the premise of existing contrasts between large-scale corporatized spheres of distribution and alternative ones.

At first glance, the distribution, marketing, and retailing of organic food in Israel seems to fit this premise. During the last two decades—after the formative years of Israeli organic agriculture and organic food production in Israel, in the Gaza Strip, and in the West Bank (as described in Chapters 1 and 2)—a rise in local demand for organic food has developed in Israel. Consequently, both large-scale and corporatized distribution practices as well as small-scale organic food provisioning have emerged. The formation of such a distribution structure for organic food in Israel may raise critical claims and questions that were directed toward organic food distribution structures elsewhere. I do not intend to reproduce the aforementioned claims about the corporatization of organic or merely describe the fragmentation of organic food provisioning in the Israeli organic field. Rather, in this chapter I intend to uncover the local-cultural meanings of this divided distribution and marketing structure and the ways in which global modes of organic food distribution have been vernacularized in the Israeli context.

Sociological accounts, such as those of Karl Polanyi, Mark Granovetter, Viviana Zelizer, Sharon Zukin, and Paul DiMaggio, have demonstrated that economic institutions as well as economic practices always have a historical basis and are always embedded in a specific structure of social relations or shared collective understandings.[13] Based on these accounts, I contend that local sociohistorical and cultural conditions have an effect on, and reflect upon, organic food distribution structure and thus deserve scrutiny. Drawing on the above-mentioned claim by Marx that "distribution should be understood as a moment of production," I ask: What social meanings are produced during processes of organic food distribution? How do sites of organic food marketing reflect and intersect with local social conditions? How do dominant actors operating in sites and institutions of organic food distribution in Israel frame the notion of "organic" and deliver this notion to their consumers? How have they branded themselves? Are these branding and marketing processes intertwined with economic and cultural changes in the Israeli food system at large, and if so, how? In what ways do the *global* social and environmental meanings embedded in the notion "organic" correspond with *local* meanings, and how is this global-local interlacing expressed in spheres of organic food distribution? These questions guided the observations and interviews I conducted in sites where organic food was packaged, delivered, marketed, and sold, and direct the logic of this chapter.

As the following description unfolds, the blurring of social and economic boundaries between two prominent socio-spatial spheres in Israeli society—the "the city" and the "kibbutz," as well as the remaining tensions and cultural differences between these spheres[14]—dictates the framing of organic food distribution processes in Israel. By exploring the first organic retail chain that operated on a national scale (Eden Teva Market) and the first collective community engaged in organic food distribution (Kibbutz Harduf) as case studies, I expose the frames and mechanisms through which the local structure of organic food distribution operates. I reveal how practices and discourses in organic food distribution in Israel are entangled with (1) the rise of Americanization in Israeli society;[15] (2) the expansion of "suburban lifestyle" and the enhancement of consumer culture;[16] (3) changes in the collective-agrarian ethos associated with the Kibbutz Movement (a formal organization and the largest settlement movement for kibbutzim in Israel); and (4) the proliferation of New Age culture among the middle and upper classes of Israeli society.[17] I also argue that the sociocultural background and the global habitus[18] acquired by prominent organic food distributers manifest in their marketing practices and in their efforts in designing and operating spheres of organic food distribution in Israel.

## Organic Supermarkets and the City

When I interviewed Moshe, a former organic farmer and a director of one of the companies that import, pack, and distribute organic food in Israel, he described some of the changes that had taken place in the Israeli organic food field since its inception:

> Today, retailers lead the organic market [in Israel]. This is true not only for Israel. Go anywhere in the world and you'll see the same picture. Ask consumers in the United States what "organic food" is, ask them, "What is the first thing that comes to mind when you hear the word organic," and they will tell you: Whole Foods [referring to the North American grocery retail chain Whole Foods Market]. And not only in America. Go to Europe—to France, Germany, Scandinavia—each country has its own "Eden" [referring to the Israeli grocery retail chain Eden Teva Market]. The organic supermarkets [*superim organim*, i.e., grocery retail chains known for their variety and selection of organic food] are the strongest players in the field.

I was intrigued by Moshe's claims. Indeed, during my research for this book, I met many actors operating in the Israeli field of organic food, and almost all my interlocutors referred, in one way or another, to the grocery retail chain Eden Teva Market.

Eden Teva Market first opened in 2003 and operated until 2019. The name of this grocery retail chain is a hybrid of Hebrew and English words that literally means "heavenly natural market." Nevertheless, the chain was the first in Israel to brand itself as a specifically organic supermarket. "Eden" (as the chain is called by its consumers) was not the first corporate retail chain to sell organic food in Israel. Nitzat Haduvdevan (literally: The Cherry Bud, a large natural and organic food grocery retail chain) opened its first branch in 1986, and in 2019, the chain had 40 stores throughout Israel. Teva Castel, another natural and organic food grocery retail chain that also specializes in nutritional supplements, was established in 1999 and now operates nine stores. But Eden was the first to explicitly utilize the term "organic" in mass media advertising. Eden accelerated the process of turning organic into "corporate organic" and put organic food at the center of the broader Israeli food system and the culinary field. Before Eden, the very few stores and retail chains that sold organic produce labeled themselves merely as natural food or health stores, even if they also offered organic food.[19]

It is hard not to notice—both because of the use of the word "market" in the name and because of the design of the stores—that Eden's operators were heavily influenced by the American supermarket chain Whole Foods Market.[20] But similarities between Eden and the American corporate organic retailer go far beyond the name. I argue that these similarities reflect a specific process of Americanization, defined as "the importation of 'American' products, images, technologies and practices by non-Americans."[21] Cindy Katz, an ex–New Yorker who recently moved from Brooklyn to Israel, echoes this argument in a short piece she published in *The Former*'s website. In her article, "Eden Teva: Israel's Answer to Whole Foods,"[22] Katz writes,

> Before my husband and I moved to Israel from Brownstone Brooklyn nearly two years ago, one of the big questions on my mind was where I would shop. Would I be able to find my staples like miso, rice paper, and quinoa? And what about organic? . . . Israel was half a world away from the familiarity of our beloved Park Slope Food Coop where we did most of our shopping, and Trader Joe's, where we did most of the rest. After a bit of time [since she and her husband have moved to Israel], we visited the branch [of Eden Teva Market] closest to us, in Kfar Saba. . . . It was such a happy day for me, and one of the first moments since moving to Israel when I felt like this enigmatic Mediterranean country really could become home. . . . Eden Teva Market is reminiscent of Whole Foods [Market].

Eden Teva Market, as described below, reflects and reproduces the Americanization of food and culture in Israel on several levels: cultural, through the American habitus of supermarket leadership; organizational, evident in the design and operation of this corporate organic chain (as well as other similar Israeli corporate organic retail chains); and socio-structural, reflected in the position of Eden and the like in the broader globalized and neoliberalized Israeli food system.

The story of Eden Teva Market's founder, Guy Provizor, exemplifies the cultural Americanization of organic in Israel. Upon his discharge from military service as an Armor officer in 1982, when he was in his mid-twenties, Provizor moved to San Francisco, where he began working in an electronics store. After a while, he bought a travel goods store, and later he even became a franchisee of vitamins and nutritional supplements. Eden Teva Market was conceived according to the models he knew from the United States.[23] The personal story of Provizor, which is similar to those of founders and CEOs of other Israeli organic supermarkets I interviewed, often appears in newspaper articles in the general and the business press. Similar to their American counterparts—such as John Mackey, the CEO of Whole Foods Market, who operates under broad public and media exposure[24]—Provizor and his counterparts endeavor to present themselves as the "face" of their corporation. In this way, their own personal image—the image of the Israeli entrepreneur who broke the boundaries of the local culture and market and discovered the "American method of healthy living" (as Avi, a CEO of another organic supermarket chain, put it when I interviewed him)—acts to identify those who engage with organic food retailing with "American success."

But the Americanization of organic food that is implied by these organic supermarkets and in the personas of their CEOs is also deeply connected to the broader organizational and sociohistorical meanings attributed to "the supermarket" as an institutional concept in Israeli society and Israeli urban culture, and I now turn to those. The first grocery retail chain established in Israel comprised grocery stores owned by the Labor Federation of the Histadrut (the General Organization of Workers in Israel) and operated by the consumer cooperative Hamashbir (known also as Hamashbir Hamerkazi, which is historically the main wholesale supplier for consumers' cooperatives and labor settlements in Israel). Since the 1920s, Hamashbir developed the consumers' cooperative movement and opened large stores that used the collective purchasing power of its members to purchase products wholesale and sell them to the association members at prices that were lower than market price. In 1932, these cooperatives were combined and served as the basis for the development of the grocery chain Co-Op (later renamed Blue Square Israel, Alon Holdings in Blue Square, and Alon Blue Square). The second important grocery retail chain in Israel was Shufersal, whose first store was opened in 1958 in Tel Aviv by Jewish entrepreneurs of North American origin and financed by US investors and shareholders. This store, located on Ben Yehuda Street in Tel Aviv, was considered "the first Israeli supermarket." Shufersal, Co-Op, and other retail chains were dubbed "self-service food chains" by mass media; they were conceived as hubs of American consumerism and have signified the increasing influence of American and consumer culture on everyday life in Israel.[25] Before the advent of supermarkets in Israel, small family-owned grocery stores and open-air markets were popular shopping venues, but for the last several decades supermarket chains have played a major role in the Israeli food system and now serve as the main places for grocery shopping.

Recently, however, various statements condemning the practice of shopping in supermarkets and large big-box stores have begun to be heard in many places, including Israel. In the United States, Michael Pollan famously urged his readers to "get out of the supermarket whenever possible."[26] This imperative was also echoed in Israeli mass media and in public discourse relating to nutrition and food choices, and several Israeli journalists and activists have sought to expose the gloomy reality and the harmful consequences of the modus operandi of the conventional grocery retail chains: environmental pollution, exploitation of employees as well as of the innocence or the ignorance of consumers, and other issues. These condemnations were intended to motivate various consumers, producers, and retailers into changing their practices. Several entrepreneurs, including the founders of organic supermarket chains in Israel, adopted this sort of discourse of "change." As such, they followed supermarket chains from the Global North (such as Whole Foods Mar-

Figure 4.1. Eden Teva Market, Tel Aviv. Photo by the author, November 2009.

ket and Trader Joe's) and framed their business operation in response to the global-structural and ideological tension between global capitalism's relentless treadmill of production and the increasing public awareness of social and ecological consequences of high-consumption lifestyles and the operation of the corporate marketplace.[27]

Eden Teva Market seemed to be the most vocal chain that adopted the rhetoric of "change." For example, one advertisement claimed: "Eden Teva Market has set for itself the goal of leading the *health revolution* in every household in Israel."[28] The chain later announced that it had initiated other "revolutions," including a "digital and e-commerce revolution," a "price revolution," and even a "dairy revolution." Contrary to the declared rhetoric of revolution and change, in the following sections I will show how Eden, like other "organic supermarkets" in Israel, gained its position in the field of organic food by using conventional and Americanized operating processes and marketing strategies while catering to the demands of middle- and upper-middle-class Israeli consumers.

## Organic Capital: Ownership, Marketing, and Branding

Between 2003 and 2015, Eden Teva Market's sales gradually increased. Their store in Netanya (a city in the northern Central District of Israel) was dubbed "the largest health food store in the Middle East." Two years after the opening of the first store, the chain's sales turnover (the total amount of revenue) was estimated at ₪24 million. Four years later, the chain's sales turnover was estimated at ₪200 million. In 2010, the chain's sales turnover was ₪300 million, and in 2014, ₪405 million.[29] These were the golden years of the organic supermarkets in Israel. The increase in Eden Teva Market's sales, as well as the relative flourishing of other organic supermarkets, would not have happened if holding companies (which also control most conventional food corporations and retailers in Israel) had not purchased controlling interests in them. For instance, Eden Teva Market was acquired in 2007 by Alon Blue Square. The holding company Alon Blue Square continued to hold a 51% value of the shares of Eden Teva Market until it was sold to Tiv Taam Networks in 2015.[30] Another example from organic food retail in Israel is the supermarket chain Organic Market (which operated under its English language name and with its logo only in English), which was established in the early 2000s by the multinational natural foods company Organic India (a company most known for its organic teas, which are sold in India, the US, Canada, and the UK) and was acquired in 2011 by the conventional Shufersal grocery retail chain (then owned by the IDB holding company). Shufersal, as mentioned above, is the same chain whose first store is considered the first Israeli supermarket and, as of 2019, is the largest supermarket chain in Israel.

As mentioned earlier, similar processes, which are often described in structural assertions—such as "the co-optation of organic by conventional food industries" or "the conventionalization of organic"—are not unique to Israel and have already been documented in many places in the Global North.[31] These cases, of the co-optation of organic movements, can be seen as instances that reflect "the new spirit of capitalism," as described by French scholars Luc Boltanski and Ève Chiapello, namely the capability of contemporary capitalism to incorporate critiques.[32] Nevertheless, this theoretical account fails to address contextual nuances, local meanings, and symbolic frameworks through which this process of incorporating critiques took place. Considering the "supermarketization" of organic food according to the understanding that markets are embedded in social contexts (and thus should be understood "in terms of their own internal dynamics"[33]) reveals that the emergence of Israeli "organic supermarkets" is part of, and reflects, the Americanization of Israel, the opening of urban Israeli consumers to global culture, and the neoliberalization of Israeli economy.[34]

Organic retail chains in Israel have embraced the marketing and management methods that characterize conventional corporate retail chains by (1) using their purchasing power as wholesale buyers of groceries by bargaining with suppliers or exerting their influence over organic food producers and suppliers to lower their prices; (2) requiring specific production and packaging practices; (3) launching private branding of organic products; (4) using financial credit to offer consumers lower prices than small stores can; (5) launching special sales; (6) establishing consumer club membership and special discounts; and (7) receiving payment for shelf space and promotional fees from large food manufacturers. Ultimately, these methods expedited the closing of small natural foods stores and independent organic food shops that could not compete with the new organic food retail chains. These methods and practices also advanced consolidation in the Israeli organic field, through which organic supermarkets merged with, and took over, local and independent shops. In addition, the branding process of these chains was done by three main mechanisms: first, by emphasizing meanings of individual health and culinary quality and by tying those meanings to global-American culinary trends; second, by changing common perceptions, beliefs, and images ascribed to the notion organic and blurring the boundaries between organic, natural, and conventional foods; and third, by de-politicizing organic food and distancing it from broader Israeli ethical (social and environmental) meanings. I now turn to describe these mechanisms as they are reflected in the case of Eden Teva Market.

At first, before Eden was owned by the holding company Alon Blue Square (2007), the chain branded itself as "the organic supermarket of Israel" and emphasized its expertise and uniqueness in marketing reliable organic foods. "When you buy organic food at Eden Teva Market, you can be 100% sure that you bought a real organic product" was written in one of the advertisements published during the chain's formative years. But it soon became clear that labeling the chain as strictly organic and appealing solely to consumers invested in organic limited the commercial growth possibilities of the chain. Consequently, a marketing process targeting a wider range of consumers was launched. Thenceforth, Eden attempted to distance the notion organic from images of an alternative culinary niche, asceticism, and radicalism. Instead, the chain's marketing directors strove to embed into "organic" meanings of a fashionable and healthy lifestyle, culinary quality, and sophisticated taste. "We worked to expand our clientele and to appeal also to consumers who are less devoted to the [organic] issue," claimed Provizor, the founder of the chain.[35] Thus, instead of "the organic supermarket of Israel," the slogan that led the process of rebranding Eden was "Eden Teva Market—just like any other supermarket."

Years before the emergence of organic supermarkets, Western lifestyle trends that focused on personal well-being, fostering dietary health and the moral valence of being a "good eater,"[36] had proliferated among many groups in Israel. Set in this context, the Israeli organic supermarkets, especially Eden, used this discourse of fashionable health food culture in order to explicitly make organic food a new "cool" culinary category. Eden regularly held "healthy and delicious" cooking workshops at the chain's branches. I attended several of these workshops where the performance of skillful chefs (some of whom branded themselves, or were branded by Eden, as "a healthy-organic chef") promised the participants that "they would share all their secrets about how delicious healthy and organic food can be" and proceeded by adding to their discourse symbolic aspects of health, care of the Self, and cooking as a leisure activity and personal wellness.

In this regard, it should be noted that before the rebranding of Eden, organic food in Israel was associated with images of dusty and not-always-fresh produce, of fruits and vegetables with wilting or rotting leaves—often riddled with worms and insects—offered to a small group of devoutly organic consumers in small natural food stores. In contrast, Eden Teva was the first organic supermarket in Israel to engage with a new, American concept of store design and organic food marketing, a concept Michael Pollan dubbed "supermarket pastoral."[37] Thus, while dedicating massive efforts to create a pleasant shopping experience and following the model of North American chains known for their organic food selections, Eden started to shift the image of organic food in Israel. Israeli Journalists Shiri Katz and Aviv Lavie (see also Chapter 5) described their first visit to one of the new stores: "A visit to Eden Teva Market provides a glimpse to the garden of Eden. After a visit [to one of the chain's stores], one can no longer associate organic food consumption with asceticism."[38] The new branding of Eden was accompanied by an increase in the selection of both organic and non-organic products offered in the store. In a newspaper interview, Provizor was asked, "Who are [Eden's] clientele?" and he replied, "As far as I'm concerned, everyone! From the organic family, those who do not bring anything home that is not organic, through naturalists who are not necessarily organic, those who consume soy milk or decaffeinated tea, and even those who had never been to a natural foods store before."[39]

In addition, the fact that Eden was designed, operated, and branded according to the models of the American supermarket chain Whole Foods Market paved the way for its integration into contemporary, Americanized Israeli consumer culture. This is evident in the location of Eden branches as well as the locations of other organic supermarkets. They could be found in shopping malls and large shopping centers, which have become central recreational places for the middle class in Israel. Even the chain's name (Eden Teva Market)—a cross between the English and Hebrew languages—is in line with the names of global and Israeli major retail chains located in shopping centers that have non-Hebrew or Hebrew-English hybrid names (McDonald's, Home Center, Shufersal Deal, etc.). Thus, marketed in sites that might be best described as "cathedrals of consumption" (places of hyper-consumption with massive sizes that enchant many consumers, to borrow from George Ritzer[40]) and conveying American images, organic food became one component of the new Israeli-American-influenced consumer culture.

This new branding of organic food was carried out while obscuring any lingering representations of political, alternative, or subversive meanings that had historically been attributed to the notion of "organic."

Eden, as well as other organic supermarkets, put a lot of effort into separating the idea of organic from ethical aspects and from similes of citizenship[41] and acted decisively to depoliticize this notion. For instance, it is notable that within these spaces one cannot find any organic food products that are also certified as fair trade. These products are available in other outlets in Israel, such as small private stores and on the websites of small producers (as well as in Whole Foods Market in the United States or Tesco in the United Kingdom). This gap is not for lack of social mobilization in Israel, for there are several associations and social movements operating in Israel that are committed to fair trade principles. Among them is Sindyanna of Galilee, a female-led nonprofit organization that promotes the sale of Arab producers' olive oil and other products (organic and non-organic) in the international marketplace. This organization operates according to fair trade principles and uses the profits to promote education and economic opportunities for Arab women. Another is Green Action (Peula Yeruka), which was a non-profit and non-governmental organization that operated between 1994 and 2004 and aimed to promote socio-ecological change. Among their activities, Green Action tried to promote fair trade and ethical consumption in Israel and thus applied the global principles of the World Fair Trade Organization (WFTO, formerly known as IFAT, International Federation of Alternative Traders) to local Israeli-Palestinians. In order to assist Palestinian food producers, Green Action branded organic and fair-trade products under the brand name Saha (wordplay between Arabic and Hebrew, as Saha is the acronym for the words "fair trade" in Hebrew and phonetically sounds like the Arabic word "health"). Eli, one of the directors of Green Action, told me when I interviewed him,

> We turned to one of the organic grocery chains. We brought with us baked goods, honey, spices, and olive oil. We even brought with us organic grape molasses, which is a very special food made from black grapes grown in small family vineyards in Wadi Fukin [a Palestinian village in the West Bank], you know . . . these families, they cannot trade fresh grapes, so the molasses is their only way to make a living from the grapes. Anyway, [he mentioned the name of an organic retail chain] they said that our products are of high quality, and they even liked the grape molasses, we even reached the negotiations stage with them. But they asked for two things we could not do: lower the price, which would have hurt the producers,

and rebrand our products. I cannot tell you for sure that the following was the reason, but that's what I believe. They did not want to sell our products with the logo in Arabic and with stories about farmers in the West Bank.

In addition to the absence of local fair-trade products, any reference to the labor conditions of the chain's employees and suppliers, statements in favor of animal rights or about preventing animal abuse, or any other ethical aspects are all lacking from the discourse promoted by these chains in their advertisement campaigns and branding processes. On the contrary, these "organic supermarkets" are packed with industrial mass-produced foodstuffs as well as imported organic and non-organic products. In accordance with marketing strategies that are common in conventional and big-box grocery retail chains—which strive to display a great deal of supply, a variety of products, and choice—Eden's campaign included a focus on its wide range of products as well as its large number of imported organic products. For instance, an advertising flyer distributed to Eden's consumers included the following: "The range of organic products we import from abroad is growing yearly. . . . Due to wide-scale import, almost everything that exists in the conventional market can be found in the organic . . . Eden Teva Market, the organic supermarket that brings you all the organic abundance that exists and changed the face of Israeli consumer culture." The emphasis on imported foods is in sharp contrast to the partial historical relation of "local-ness" to the idea of organic.

This was particularly evident in the fruit and vegetable department in Eden's stores. In the early stages of my field work (2008), I noticed signs bearing the names of the farmers who grew the produce (all of them Israeli farmers) and the places where the fruits and vegetables were grown (various locales throughout Israel). However, over the years I noticed the gradual removal of these signs, until they disappeared altogether. It seems that in the area of marketing fruits and vegetables, the Israeli organic supermarkets adopted the practice of the "fetishization of fresh food," which is prevalent in conventional retailing.[42] Marxist geographer David Harvey claimed that "the grapes that sit upon the supermarket shelves are mute; we cannot see the fingerprints of exploitation upon them or tell immediately what part of the world they are from."[43] Apparently, organic fruits and vegetables in the Israeli "organic supermarkets" are also mute.

Thus, instead of evoking ethical meanings, Eden's campaign was focused on the wellness and self-care of its individual consumers. In a series of advertisements, numerous Israeli celebrities, who were known for being active in sports, having notable weight loss, or looking young for their age, were depicted on Eden's advertisements declaring, "I started to take care of myself. Eating healthy? Me too!" This coupling of the notions of organic and self-care was also promoted by the giveaway of tickets for spa treatments, discounts for membership to fitness centers, and the like. In addition, workshops and lectures held at the supermarkets contributed to the association between organic and individual wellness. I attended some of these workshops, whose titles testify to the intention to equate organic food with a healthy lifestyle: "Nutrition and Skin Care for Menopause" and "Dietary Supplements—What Should You Take?"

Eden's campaign and marketing strategies were among a growing number of branding efforts—by both conventional grocery retailers and organic retailers in Israel—that harnessed the global proliferation of health, wellness, and self-care rhetoric. Within these dynamics, Eden contributed profoundly to the process of conventionalization of organic food in Israel and to the formation of the Israeli version of corporate organic operating inside the Israeli market. This conventionalization of organic is evident in the chain's chief advertising slogan: "In Eden Teva Market you will find a huge selection which ensures that everyone can find the exact products that suit them. Just like any other supermarket . . . only healthier!" The chain's advertising campaign also included a series of depictions and textual expressions, seeking to identify the chain with variety, quality, and health. For example, in the chain's advertisement series, which was broadcast on the radio, television, and new media platforms, Israeli celebrities declared—as if they were stunned by what they saw when they visited an organic supermarket—"Look at the variety! Look at the quality!—It's good for you!"

The range of images that emerged from this campaign worked to dissociate organic food from the asceticism and tastelessness that it connoted for many years. Consequently, when buying organic food at Eden, consumers can capitalize on this practice to gain cultural capital and to manifest their prestigious habitus,[44] which consists of conspicuous affluence, freedom of choice, special taste, and self-care.

## American Dreams:
## Organic Supermarkets and Urban Consumers

Exploring the meanings that Israeli consumers create by shopping in Eden Teva Market (as well as their experiences of organic food consumption in similar organic supermarkets in urban places) reveals that messages of health, wellness, and self-care promoted by the supermarkets are central to the qualities consumers ascribe to the notion of organic.[45] A theme that repeated itself in my interviews and engagements with consumers in organic supermarkets in Israel is a concern with broader cultural trends of Americanization and with the ways in which "neoliberal branding"[46] resonates in the everyday lives of middle-class consumers in Israel. For example, Noa, a 41-year-old woman, testified, "Since Eden opened, I became an enthusiastic patron of organic supermarkets. I was fed up with depressing health stores, and I do not like the vegetable box scheme [CSA]. I don't like boxes of bad vegetables sent to my home. It's not for me. I invest a lot of time and money in cooking; it's my hobby. I host a lot and plan ahead of time what to cook. I can't have farmers bring me a box with what they just grew and not with what I need to cook. . . . I enjoy shopping and I want to be able to pick and choose the best product. Variety, that's the big benefit of Eden."

Indeed, the stores of the chain had a huge selection of food products, both organic and non-organic. A tour in Eden conveyed the sense of a visit to a "global cultural supermarket"[47] with its dozens of legumes, spices, and dried fruits imported from around the globe, tea infusions and a selection of imported coffee beans, gourmet halva (sweets made of sesame and sugar) from Turkey and the West Bank, food products from the Far East, Europe, and Central America, and much more. Similar to the experience of shopping in American retail stores, such as Trader Joe's or Whole Foods Market,[48] shopping in Eden Teva Market seems to be a practice that constantly engages with the "seductiveness of variety."[49] All consumers interviewed—those who visited Eden Teva Market regularly—referred to the wide range of products and their quality. Many of them indicated familiarity with American retail chains and described their experience in Eden as "a short trip abroad." For example, Orit, a 41-year-old woman from Tel Aviv, said, "In Eden, I feel as if I'm suddenly abroad, just like being abroad once a week! This is the only supermarket where I really feel that I have a variety of interesting and surprising things . . . and the fact that it's organic is just obvious. That's how it is today in America and Europe. See what happens in Trader Joe's or Whole Foods. Go to the US and you'll see—the best supermarkets there are the organic supermarkets!"

Many consumers appreciated Eden for adopting administration and marketing methods similar to American discount stores (which are widespread in the conventional retail food sector in Israel), and thus they often credited the chain for contributing to the (relative) decline in prices of organic food in Israel and increase in the supply. Some consumers maintained that the emergence of the chain and its operation in the retail food sector in Israel was a matter of utmost importance. According to them, the importance stems not necessarily from the chain promoting environmental or social values but rather from its hand in opening the organic food sector to free markets and free trade. I found a neoliberal consumer discourse to be common among Eden's consumers. The words of Baruch, a 50-year-old engineer whom I met on one of my visits to an Eden store, represent this neoliberal perspective clearly: "I have no doubt that Eden Teva Market did a good thing. . . . 15 or 20 years ago, we bought two tomatoes, a cucumber, and an apple and paid 200 NIS. Today I sometimes find organic [food] cheaper than non-organic—so what's wrong with that? It's the same also in America. We lived [pointing to his wife] for a while in Palo Alto. There you can get organic [food] at reasonable prices. Now it is also happening in Israel. The competition [between Eden Teva Market and other organic supermarket chains] lowered the prices." Baruch sees the way Eden Teva Market operates as an expression of the "American way"; that is, promoting a pleasurable experience and cheaper organic food that supposedly arises from capitalist competition. Furthermore, there are some who consider consumption in Eden Teva Market to be a practice that contributes to society and the environment. For example, Yoav, a 38-year-old man from Tel Aviv, said, "I don't shop here every week, it's a bit expensive. But listen—there is awareness and care [in the way the chain operates]. So even if I spend a little more [money], at least I'm doing something for the environment." Yoav's words illustrate the assimilation of environmental messages and images seen in Eden Teva Market, which are expressed in the multiplicity of shades of green, in the fabric bags in which the products are packed, in the recycled paper packing for baked goods, and in the verbal and visual signs on the walls of the supermarket referring to "nature" and "environment."

However, I found a lack of consideration among customers for the negative socio-environmental consequences of the chain's operation—the consolidation promoted by the chain within the field of organic food and organic agriculture, the environmental costs inherent to the operation of a big grocery retail chain such as Eden, labor conditions, or the deepening of inequality between citizens through inequitable access to healthy food products due to the cultural and economic capital required to shop there. Even those who did criticize the chain presented a soft attitude. For example, Dikla, an organic food consumer and naturopathic practitioner, said, "I know all about the problems of 'Eden' and agree that it is a profitable, industrial business. But it is important to see the good sides. It is not a conventional supermarket. No! It's an organic supermarket, to a certain extent. Speaking of organic—there are many missing things here: this is not a cooperative, I know. Nothing here is communal, of course. It is a business, after all. . . . True, they're interested in profit, but at least there is healthy food here, and thanks to them the prices of healthy food are no longer skyrocketing. You know what? That's why it works. I would not be able to buy organic today if not for Eden. I used to try to force myself to go to natural food stores. But it was impossible. Neither the price nor the quality. . . . Unfortunately, everything is becoming chains. That's how it is all over the world. That does not make me happy, but if that's the case, then at least there is one chain where you can find healthy organic food at reasonable prices. It's something, too, isn't it?"

To summarize, the translation of the term "organic" by the urban consumers of Eden Teva Market, as well as by its directors, designers, and managers, is composed of predominantly consumerist patterns and access to global material and cultural goods. In the course of symbolic consumption of "organic," Eden's consumers celebrate the expansion of their access to global taste cultures. The health representations connected to the notion organic express individual meanings, healthy lifestyles, trendy-Western/American lifestyles, and up-to-date culinary style. Thus, while prioritizing consumer desires over citizenship ideals,[50] Israeli corporate organic consumers cultivate a cosmopolitan identity. When ethical meanings are attributed to their consumption practices, they are presented according to the paradigm of the neoliberal green consumption as well as the Western trend of commodification of environmental and health concerns.

In parallel to the growth in the volume of organic food sales in urban places, organic food distribution and marketing has been developed in another important socio-spatial entity in Israel: the kibbutz. In an essay titled "Others," historian Ze'ev Tzahor described that "the city" and "the kibbutz"—which were founded, symbolically, on dates close to each other (Tel Aviv, "the first all-Jewish city in modern times," and Degania, the "mother of the kibbutzim," known as the first socialist-Zionist farming commune in the Land of Israel, were established between 1909 and 1910)—became two of the most important socio-spatial institutions of the Jewish-Hebrew culture and society in the Land of Israel. The city symbolized the ethos of cosmopolitanism and the connection to the world ("to be a nation like all others"). The kibbutz was considered a social avant-garde initiative of the Jewish people, who sought to establish a Jewish society on revolutionary social foundations ("a light unto the nations"), and was perceived as the crowning achievement of the Jewish settlement movement.[51] However, as described below and as revealed from looking at places and practices of organic food distribution in the kibbutz, the common denominator is greater than the difference between "the city" and "the kibbutz."

## Organic Kibbutz

In a newspaper article titled "No Chemistry: Organic Food for the Masses," Aviv Lavie[52]—an Israeli journalist writing on environmental matters—distinguishes between two common approaches to organic food and agriculture:

> The world of organic is divided between two approaches: one which determines that in order to compete with [the conventional] food industry, the shelves in our stores must serve as a sort of "organic mirror," on which one can find an organic version of every conventional food—processed and industrialized as it may be. The second approach [argues that] organic culture should provide a real alternative, and concentrate on growing and marketing fresh, unprocessed, and non-industrialized foods.[53]

Expressions of these two approaches—the "organic mirror" approach and the "alternative" approach—are both evident in Kibbutz Harduf, a communal settlement located in the lower Galilee in northern Israel. The entrance to Kibbutz Harduf is surrounded by oleander plants, for which the kibbutz is named.[54] From the entrance to the kibbutz, one can see the business offices of Harduf Organic Food Products (HOFP) and an array of industrial food processing buildings surrounded by trucks from which people unload raw materials for the packaged industrial food products produced in the factories. When I first visited Kibbutz Harduf in 2011, HOFP was controlled by Tnuva, one of Israel's largest food corporations. Signs with Tnuva's well-known logo were prominently displayed all around the kibbutz and indicated that HOFP seems to model the "organic mirror" approach.

Just a short walk from the industrial zone, one arrives at the heart of the kibbutz, a plaza where craft workshops and small initiatives operate. Among them are an organic restaurant, a garden, and what is possibly Israel's largest local organic vegetable and fruit warehouse. Not far from the plaza one can see an organic dairy and a cowshed. The waste from the cowshed provides compost for the agrarian initiative in the kibbutz and the fields around it. The heart of the kibbutz, therefore, reflects an expression of the "organic alternative" approach. The two approaches operating next to each other in Kibbutz Harduf seem to contradict one another. However, as described below, a closer look at the industrialized organic food production plant and at the small-scale initiatives operating side by side reveals some of the common denominators between the two and tells us about their shared cultural logic.

The case of organic food in Harduf exposes how changes in the communal ethos attributed to kibbutzim, as well as changes in their agrarian ethos and broader changes in the cultural and economic structure of Israel's Kibbutz Movement, affected and reflected on the role of kibbutzim in the conventionalization of organic food distribution in Israel. At the same time, the case of Harduf reveals how "alternative" ideas—such as New Age spiritual ideas—were attached to the meanings of the notion "organic" as well as to the processes of marketing and distributing organic food. Ultimately, these mechanisms of distribution contributed to the commodification of organic food at large.

The production and distribution of organic food in Harduf are entwined with the main sources of income that are customary in kibbutzim in Israel: agriculture, industry, tourism, and hospitality.[55] Traditionally the backbone of Israel's agriculture, kibbutzim nowadays produce 40% of the country's agricultural crops and 7% of the industrial output, and they account for 10% of the tourism market.[56] HOFP is one of about 250 industrial plants operating in kibbutzim (of which about 50 are food factories),[57] and Harduf Restaurant, which serves organic food to tourists and travelers in the Galilee, is one of the many recreational, culinary, and hospitality attractions flourishing in kibbutzim. Still, as anthropological research on kibbutzim reveals,[58] each kibbutz is different. Therefore, while Kibbutz Harduf may seem similar to others, it has its own uniqueness.

Harduf can be seen as a hybrid space in which the modern utopian ethos of the Kibbutz Movement—communality, egalitarianism, and agrarianism—is intertwined with postmodern, New Age culture. The former—the ethos of the Kibbutz Movement—refers to the model of the kibbutz in Israel: a self-contained social unit inhabited by Jewish members (*kibbutznikim*) where property and means of production are communally owned. Following the kibbutz financial crisis of the 1980s, the Kibbutz Movement in Israel faced socioeconomic decline and a collapse of its ideological tenets. After this economic and social crisis, many of the kibbutzim were "privatized," meaning they downsized their communal institutions.[59] Kibbutz Harduf is one of the 80% of all kibbutzim in Israel that were privatized (as of 2018).[60] In addition, Harduf can be classified as an *alternative space*, as anthropologist Dalit Schimai puts it: a settlement whose residents hold a strong attachment to the cultural practices associated with New Age culture.[61] These spaces have ample activities aiming to promote spiritual development and transformation that foster both "personal and cosmic harmony."[62]

Scholars indicate that New Age culture has emerged in the context of a Western, globalized consumer culture and expresses the cultural logic of late capitalism, such as individualism, entrepreneurism, and freedom of choice.[63] In the Israeli context, it has been indicated that New Age spiritual and body practices—with their theological focus on individual welfare and self-realization—are common among privileged groups and form part of the contemporary new middle-class lifestyle in Israel.[64] Interestingly, the main sociological explanation for the growing popularity of New Age culture in Israel is somewhat similar to the explanation for the change the kibbutz underwent: the transition of Israeli society from a collectivist ethos to a multi-sectorial and individualist one.[65] In the case of New Age culture, this transformation is manifested in the adoption of post-materialist values, in the search for individual fulfillment, and in the realization of the growing impact of global processes.[66]

Kibbutz Harduf was established in 1982 by a group of New Agers, the sons and daughters of the Zionist Jewish settlers in Israel (*halutzim*), all of European descent (Ashkenazi), well educated, secular, and born and raised in other kibbutzim. Prior to establishing Harduf, the founders traveled to Europe, where they were introduced to, and captivated by, the anthroposophical theories of Austrian philosopher Rudolf Steiner (1861–1925).[67] Upon returning to Israel, they sought to form a cooperative community that would live according to the spiritual science and lifestyle of anthroposophy. After applying to the Kibbutz Movement with the request to establish a new kibbutz, they were allocated lands and started to farm the fields near Zippori (Northern Israel). These lands were state lands within the jurisdiction of the Jezreel Valley (Emek Yizrael), but they were difficult to cultivate. For that reason, and presumably due to the intention to advance the project of "Judaization-dispersal policy," which aimed at achieving a demographic balance in favor of Jews in the Galilee,[68] the lands of Kibbutz Harduf were given quite willingly to these idealistic young settlers. In the eyes of the Kibbutz Movement, this community was considered somewhat eccentric but still worthy of support since it included former kibbutzniks. Over the years, initiatives in the fields of education, therapy, alternative medicine, arts, and biodynamic agriculture were developed at Kibbutz Harduf and served as sources of income for its members. In line with the Israeli version of postmodern and New Age discourse, many of these initiatives presented themselves as dealing with "renewal" (*tikun*) and "healing" (*ripoui*).

For nearly two decades, organic food was produced in Harduf for the subsistence of the kibbutz residents and as part of small-scale marketing operations. These food production activities were organized under what was then a small company simply branded Harduf and owned by the members of the kibbutz. The founders of the organic agricultural projects in Harduf tended to frame their actions in terms of both ethical values and idealistic motives. Guy, who served as the first CEO of Harduf, says, "People from the Kibbutz Movement considered organic agriculture to be a strange thing. They did not see it as practical, especially from an economic point of view, because there was no market for it. They did not understand that our reference point was not economic but rather based on principles."[69] When the demand for organic food products in Israel increased, the kibbutz members realized that organic food had economic value in addition to the moral values they had ascribed to it. The kibbutz became a vanguard on many fronts of organic food production, including dairy farming, composting, wheat production, flour milling, pest control, and preparing artisanal organic foods. However, those who were occupied in organic food production in Harduf were constantly struggling to efficiently transport and distribute their products.

During the late 1990s, the agri-food enterprise Tnuva expressed an interest in purchasing the agricultural production facilities in Harduf, particularly the organic dairy farm and the organic dairy production initiative (which was then—and continues to be—the only organic dairy in Israel). In 2002, Tnuva acquired the kibbutz's production facilities and the rights to use the brand. The consequence of this acquisition can be seen as the first case of conventionalization and diversification (post-Fordism) of organic food in Israel. Soon, the brand Harduf changed from being a marginal, unknown biodynamic and organic food producer to a well-known, respected brand among Israeli consumers. Harduf became so well known and so strongly associated with organic agriculture in Israel that for Israeli consumers, even those who are not regular consumers of organic food, the term "organic" is inextricably associated with the brand Harduf. Under Tnuva's management, the production at Harduf (HOFP) expanded to include products grown on farms outside the kibbutz (organic breakfast cereal, organic ketchup, and frozen chicken meat) and imported products (frozen organic beef from Argentina and organic pasta from Italy), which were sold by Tnuva under the brand Harduf.

In addition to the prominence of Harduf's logo on the foods pro-duced, packaged, and distributed at HOFP, many of the food packages also included illustrations of the Galilee landscape where Kibbutz Harduf is located. These foods were branded and marketed as "organic food from the Galilee." Thus, the products were immersed in symbolic-spatial properties and depictions of (imagined) northern-rural Israel—whether they were produced on the kibbutz or only imported and packaged there.

Following Tnuva's acquisition, agricultural projects in Kibbutz Harduf—whose primary purpose, according to the founders, was to foster anthroposophical values and provide economic independence and liveli-hood for kibbutz members—were pushed aside. Therefore, many of Kibbutz

Figure 4.2. Flyers—"Harduf (HOFP) brand products": conventionalization of organic food in Israel. Photo by the author, December 2019.

Harduf's organic and biodynamic initiatives shifted to market-oriented ones. The firm (HOFP) became preoccupied with maximizing economic profits, developing cutting-edge production technology (especially in the area of the organic dairy), marketing processed foods labeled as "organic certified" on a national scale, and engaging in global trade (exporting commodities made in the kibbutz and importing agricultural inputs). Thus, the commodification and conventionalization of organic became a core activity in Kibbutz Harduf.

Concurrently, small-scale organic food production and distribution initiatives—such as Harduf Restaurant, Harduf Community Garden, and a store selling local organic fruits and vegetables—began to flourish in the kibbutz (reflecting the "organic alternative" approach outlined by Lavie). Thus, one might wonder how these two supposedly contradictory approaches flourished in Kibbutz Harduf. What are the social meanings embodied in these opposing modes of food production, distribution, and marketing in Harduf? In the course of providing answers to these questions, I consider the case of Kibbutz Harduf in relation to broader social, economic, and cultural changes in Israel. Specifically, I examine how egalitarian principles (for example, those of the kibbutzim) collided with neoliberal ideology and logic,[70] how industrial market actors met with countercultural movements,[71] and how all of the above are expressed in the material and cultural production in "alternative New Age spaces" such as Kibbutz Harduf.[72] In order to answer these questions, the following empirical sections examine two sites that epitomize the split in modes of food production in Kibbutz Harduf: Harduf Restaurant and Harduf Organic Food Products (HOFP).

## Gastro-Anthroposophy and Fine Dining in a Kibbutz

Harduf Restaurant is located near the commercial center of the kibbutz. Nearly all those who visit Kibbutz Harduf also visit the commercial center, a plaza surrounded by several shops. One of the shops is the grocery store known as *tsarkhaniya*. The term *tsarkhaniya* is used throughout kibbutzim and translates to a store that sells groceries only to members of a given collective community without any profit or with low profit. In the past, during the "golden years" of the kibbutz (1930s to the late 1970s, according to most kibbutz scholars),[73] the tsarkhaniya was an important feature of daily life, second only to the kibbutz dining hall.[74]

Nowadays, *tsarkhaniyot* (grocery stores in plural)—including the one in Kibbutz Harduf—operate as regular for-profit stores frequented not only by members of the kibbutz but also by occasional visitors.

The tsarkhaniya in Harduf is equipped with sliding automatic doors with noticeable HOFP logo stickers on them. Just inside the entrance stands a cabinet containing a variety of organic food products made by HOFP. At first glance one might think that only organic produce is offered in this store. However, after passing through the entrance hall, an abundance of non-organic industrial food products, just like those sold in every conventional retail store in Israel,[75] can be found. Popular Israeli industrialized snack foods, such as Bamba (a puffed peanut snack, an Israeli version of Cheetos), Shkedei Marak (crisp mini-croutons made from flour and palm oil used as a soup accompaniment), and Crembo (chocolate-coated marshmallow) can be found on the shelves, as well as imported organic oatmeal, industrialized and imported mayonnaise, and jars of "Harduf Organic Tahini" (an organic tahini paste produced in a Jewish factory located beyond the Green Line in the West Bank, bought by HOFP, and distributed with HOFP's logo imprinted on its packaging). The refrigerator is stocked with conventional cottage cheese, and Tnuva's conventional milk cartons are prominently featured. Ironically, Harduf's organic milk, produced from cows just a few miles from the grocery store, is less prominently displayed.

Harduf Restaurant seems to be antithetical to the tsarkhaniya in Harduf. Unlike the tsarkhaniya's mixture of conventional and organic foods, Harduf Restaurant is strictly organic. One could assume that Harduf Restaurant serves as the central place where kibbutz members consume their food and perhaps even serves as the kibbutz dining room.[76] However, Harduf Restaurant has never served as the kibbutz's dining hall and the members of Kibbutz Harduf do not frequent the restaurant very often. They prefer to cook at home and buy groceries in the tsarkhaniya—nowadays the main place where food is sold in Kibbutz Harduf (and which has practically replaced the kibbutz dining hall). Most of the diners in Harduf Restaurant are tourists, volunteers, or trainees in one of numerous educational programs operating in the kibbutz.

Visiting Harduf Restaurant felt like visiting a museum, since this place serves as a center for dispensing not only organic foods and dishes but also anthroposophy culture and anthroposophy-friendly artifacts. At the entrance to the restaurant, one cannot help but notice a huge bookcase filled with books about anthroposophy and New Age culture: an impressive collection of Rudolf Steiner's philosophical writings translated into Hebrew, children's books about environmental ethics, healing and self-care books produced by small publishers, and vegetarian cookbooks. The books also include *The Living Kitchen: Organic Vegetarian Cooking for Family and Friends* by Jutka Harstein (2012). Harstein runs the restaurant. In addition to the bookcase, there are racks with women's clothing made from light fabrics, shelves loaded with handmade products from the workshops that operate in the kibbutz, and special cookware (including brands such as Le Creuset). When I asked Harstein why such expensive and exclusive brand-name iron pots were sold in her restaurant, she explained, "Anthroposophical cooking culture encourages people to pay attention not only to what they cook but also to how they cook and what cooking utensils they use. Cooking with iron is extremely important. Iron has real life in it. The origin of iron is in the soil. It is not enough just to use organic products. The pots must also be organic."

During one of my visits to the restaurant, I saw a group of local tourists arrive for a late breakfast. Harstein served the dishes while describing them with enthusiasm and pride: "Here we have almond pâté . . . try our special tofu-mayo spread. . . . This is pesto made from organic basil and parsley." Later, the diners asked her about her Hungarian-Romanian origin, and the conversation went as follows:

[Diner 1:] Did you grow up on healthy food?

[Harstein:] You will not believe what food I ate as a child. When I visited Transylvania, the place where I grew up, I recalled the foods I had grown up on. It's hard to believe that we ate those things. Now I realize it wasn't healthy food, but . . .

[Diner 2, interjecting:] Well, it was very different back then. Today it's easy to get organic food, certainly here in Harduf. I can imagine that it was not easy to get organic food in Transylvania a few decades ago, was it?

[Harstein, dismissively:] It was not vegan food, we ate a lot of pork and lard and there were hardly any fresh vegetables, but it's actually not so different from the food we have here in Harduf, if you think about it. According to my understanding, it was very much like organic food. From my current point of view, considering my nutritional knowledge, the food [that I grew up on] was not healthy at all, but it was natural, there were no chemicals in it, and most of all, I was absolutely nourished by it spiritually. The food was full of spirit, it was full of life, and this is what I would like to do here in my restaurant. So, I think that I actually grew up on organic food. To be sure, it's not like what I cook today. Now I can cook for you food that is organic, healthy, and spiritual at the same time. Food that is connected to both the earth and the spirit. OK, we talk too much . . . Please, taste it, start eating.

Perceiving food as "spirit," as evidenced in the conversation above, is a recurring motif in Harstein's discourse. For example, in her introduction to The Living Kitchen, she emphasizes repeatedly: "Food is spirit."[77] After reading her book, watching her various appearances on television cooking shows, reading interviews conducted with her in popular media, reviewing her articles in Adam Olam (a Hebrew magazine for anthroposophy and Waldorf education), and attending several of her public lectures, it is hard not to appreciate and recognize the "life forces" and the "formative forces" she sees in food. She is devoted to spreading the word on the importance of eating healthy and organic food. And she insists on giving her own definition of "organic": "In order for a certain food to be organic, in my opinion, it should be natural, chemical free, full of spirit, and made with love."

In accordance with practices shared by New Agers, which focus on "personal change" as a central means of expediting spiritual cosmic transformation,[78] Harstein constantly uses her own personal story in her talks and writings. Her story serves as her main colloquial apparatus to elucidate her understanding of the values and meanings of organic food. Harstein immigrated to Israel in 1974, lived in Kibbutz Kfar Menahem (located in southern Israel), and worked in the kitchen. Although she was not a vegetarian herself, she was in charge of preparing the vegetarian lunches for 150 kibbutz members. "To help me overcome my total ignorance of such cooking, my American mother-in-law sent me The Moosewood Cookbook,[79] then a new bestseller." Browsing the well-known American vegetarian cookbook, which is now considered an important part of the American countercultural canon,[80] made her consider food as part of a wider cosmological conception: "That book altered the way I thought about food. I became so ardent about vegetarianism that my then-husband said, 'food is not everything.' But for me, it was."[81]

During this period, she was exposed to the anthroposophical theory and became fascinated by it. In the early 1990s, she moved to England, to Forest Row in East Sussex, to study anthroposophical education. She got a job as a cook at Nutley Hall, a home for people with special needs. "We cooked good food there," she told me when I interviewed her, "although it wasn't the best. I remember I used a lot of industrialized organic processed food produced by Infinity Foods, the British 'Harduf,'" she explained, referring to HOFP, the company that was owned by Tnuva. "But we could also use some good local products from the gardens there, and these were grown by biodynamic methods." While working at Nutley Hall, she planned to open a "healthy café, a place serving good food,"[82] where she could practice the things she learned. But soon, she had to leave England because of visa issues, and a friend from Kibbutz Harduf told her, "There is a restaurant in the kibbutz that is screaming for a mother."[83] Thus, she moved to Kibbutz Harduf and stayed there. All these experiences influenced her current activities, as she writes in the introduction to *The Living Kitchen*. "If *Moosewood Cookbook* was my high school, those years in Forest Row and Nutley Hall were my university! The best that exists! The doctorate was Harduf's restaurant; this book is my dissertation."[84]

Nowadays, she proudly claims, many recognize her as "the face of Harduf": "It's important to me that people appreciate who I am. It is important to me that people know that I live in [Kibbutz] Harduf and that one can find good organic food here. I also want people to know that the restaurant here is special. People who are familiar with the brand name Harduf know that organic food is made in [Kibbutz] Harduf. And yes, Harduf [HOFP] indeed produces organic food, but this is organic food that one can buy at the supermarket. That's fine, but the organic of the supermarket is not *my* organic. The food I serve is always loaded with life forces."

In public lectures, she presents herself as an expert on "nutrition from a spiritual viewpoint." At one such lecture she said: "Our digestive system is part of a larger system. It unites humanity with earth and nature. We must understand that there is a strong connection between eating and ecology, between the sun, the food and our body, between nutrition and spirituality." Her listeners and the diners at her restaurant do not remain indifferent to the spiritual elements with which she "spices up" her dishes. For example, a consumer posted a recommendation for the restaurant in a blog that deals with culture and municipal politics in northern Israel: "If you have not yet eaten philosophy or tasted anthroposophy, Harduf Restaurant is the place for you." In this regard,

Harstein's dishes can be seen as symbolic goods which testify to her vast spiritual capital (that is to say that her spiritual knowledge, competencies, and spiritual lifestyle preferences have become positional goods within symbolic economies and cultural fields).[85] Using techniques that can be defined as spiritual capitalization (and even spiritual commodification), she constructs (and sells) her organic food not only as healthy or natural but also as spiritual food. "It is important to distinguish [between] those who do this [produce and cook organic food] out of both deep inquiry into themselves and broad spiritual perspective," she told me. "You need to figure out who prepares organic food out of his or her soul, from deep and personal drives, and not for external or economic reasons."

When I interviewed Harstein, we sat in her restaurant in front of a window overlooking the plaza ("commercial center"). It was an early afternoon in spring, and we saw many children and young adults coming back from school. They went in and out of the tsarkhaniya, treating themselves to all kinds of non-organic industrial snacks. Harstein described the restaurant as a "bubble" that operates separately from its surroundings. "The kibbutz members," she said, "do not frequent the restaurant, and in order to maintain the restaurant and myself, I have to turn to people outside the kibbutz." Her wish, she told me, is to open a restaurant in Tel Aviv. Thus, if the restaurant was her doctoral dissertation, as she put it, her aspirations for Tel Aviv could be her plans for her "academic career."

Meanwhile, she works hard to reach and appeal to a wider group of consumers and to bring her organic-spiritual food to the center of the Israeli culinary field. Every few months, she travels to Tel Aviv to cook as a "guest chef" at a fashionable café. One of the flyers distributed at the restaurant advertising one of the culinary events she organizes read, "Welcome Jutka, the chef from Kibbutz Harduf. She will cook vegan, spiritual, and Galilean food [food of the Galilee] for us. Get ready for an unforgettable organic Galilean meal." When I asked her about her work in Tel Aviv, she told me,

> I would like people in Tel Aviv to eat some really good food. I know that there my creativity will be appreciated. I see the reactions [when diners taste my] walnut terrine and my almond fritters. These delicacies are not only organic and healthy, they are tasty and special. Even the biggest carnivores praise it and say how much they enjoyed it! Why? Because it's not only organic, it is special. There's nothing like it.

While adopting practices of innovation and experimentation, such as those accepted at the high end of the restaurant industry,[86] Harstein strives to elevate her menu so that it will be conceived of as not merely "organic," "anthroposophical," or "vegetarian" but also *gourmet*. Thus, while working as a culinary consultant for fashionable cafés, she created dishes such as Eggplant Moussaka with Quinoa and Feta Cheese, Stuffed Seasonal Vegetables in Rich Thai Sauce, Roast Loaf of Nuts with Apple Sauce and Antipasti Salad, A Casserole with Colorful Vegetables and Whole Grains, Flourless Orange and Almond Cake, Carob Cake with Raisins, Nuts, and Captain Morgan's Rum, and many more. "Don't get me wrong," she told me, "when I cook there [in restaurants in Tel Aviv], I always tell the cooks: 'your customers should realize that there is logic in the salad or in any other dish.' Whatever they cook must always be connected to both spirit and earth." Thus, by combining fine dining with natural and vegetarian eating, and with spiritual images, she pursues a higher status for herself as a chef in the Israeli culinary field. Consistent with the cultural logic that has guided many New Age movements and many New Agers in Israel,[87] Harstein's culinary practices have facilitated the spread of "alternative gastronomy" within mainstream culinary culture and have enabled the assimilation of "spiritual organic" in the broader culinary field in Israel.

## "Do Not Call Us Alternative":
### Industrialization and Marketization in Kibbutz Harduf

In contrast to the Harduf Restaurant, which is immersed in spiritualism and "alternative" culture, Harduf Organic Food Products (HOFP) has aimed explicitly toward assimilation into mainstream industrial food production. When I studied HOFP through observations and conversations with directors and employees who were members of Kibbutz Harduf, it was difficult to discern signs of "alternativeness," "resistance," or "counterculture" in their practices and discourse. This reluctance to resist and avoidance of countercultural actions elucidate broad sociocultural and political economic processes: a deviation from the original model of kibbutzim (originally, a hybrid between egalitarianism and Zionism) and the reformation of kibbutzim according to meritocratic and post-Zionist ideologies. It is also exemplified in the ways in which kibbutzim and other "utopian-intentional" communities in Israel (such as moshavim) have changed due to the globalization and economic neoliberalization of Israeli society.[88]

I learned about the HOPF from Yishai Shapira, who worked at HOFP for 14 years. He was among the founders of the company in 1991 and led the initial process of conventionalization of organic food in Israel. Later, in 2004, he also directed the sale of HOFP to Tnuva and became an employee of the latter. For nine years he ran HOFP as its director general. It is remarkable how Shapira's personal biography, habitus, and sociocultural background are congruent with the "cultural biographies"[89] and stories of the companies in which he worked (HOFP and Tnuva): stories about the transition from being immersed in a collectivist ideology to taking a role in capitalistic processes driven by global market forces.

Being a *kibbutznik* (the word for a kibbutz member), Shapira can presumably be best described as affiliated with the upper echelons of the hierarchy of Israeli society, one of the "offspring of the Zionist 'founding fathers' and army leaders, the farmers of yesteryear and the fighters of yesterday."[90] He grew up and was educated in Kibbutz Ein Dor (located in the lower Galilee, northern Israel, not far from Kibbutz Harduf). Ein Dor, his birthplace, was the first Jewish settlement founded in Israel after the declaration of statehood (1948) by members of the Hashomer Hatzair (a socialist-Zionist, secular Jewish youth movement). Today Ein Dor is a privatized kibbutz, one of those that underwent a process of transition from the collectivist motto of "from each according to his ability to each according to his needs" to economic models of differential payment, privatization of means of production, and privatization of the ownership of residence.[91] These changes in ideals and values are also reflected in Shapira's personal story. During his military duty service, he was expected to be a successful officer. It was the late 1980s—a period in which economic changes, such as neoliberalization, post-industrialization, and privatization, as well as cultural-ideological changes took place.[92] Shapira, in accordance with the zeitgeist, decided not to serve the nation as an officer in the standing army and returned to work in the kibbutz. He started to work as a farmer in Kibbutz Ein Dor and was responsible for the field crops production sector, at which point he was exposed to the Western "organic ideal." When he was asked to manage the agriculture sector in Kibbutz Ein Dor, he insisted that the kibbutz should allocate lands for the cultivation of organic produce. The kibbutz members approved, and soon residential areas of Kibbutz Ein Dor were surrounded by fields where organic crops, such as cucumbers, corn, and artichokes, grew. After earning a degree in agricultural economics, he became interested in new agricultural and economic possibilities. "I identified organic agriculture as an area that had enormous potential for business, an area that I would be able to develop, and at the same time to develop myself within it," he said in response to my queries about his motivations for getting into organic agriculture. An article that interviewed him reported,

> [There were] two points in time . . . for bringing about change. The first occurred at the beginning of the 1990s, when there were initial signs of peace between Israel and its neighbors, something to which "the organic market reacted positively, because there is a connection between organic food, freedom and peace." The second took place in 2000, when the U.S. passed a law standardizing organic agriculture, which "by its very acceptance and application, caused a global push forward in the field."[93]

In 1991 he moved to Kibbutz Harduf following an offer to work in biodynamic agriculture in Harduf's organic food production sector. After a short time, he took a management position and led a process of reorganizing the biodynamic agricultural activities. He sought to coordinate and unite all farming initiatives operating in Kibbutz Harduf under one company, which would focus on production and distribution and would increase the economic profits of the kibbutz. In this respect, he worked toward a much larger scale of organic food production and distribution than had been practiced until that point. Sociologically, his actions can be best described in Weberian terms: he pursued a transition from *substantial rationality* (in this case biodynamic practices, which were based substantially on anthroposophical values) to *formal rationality* (in this case the increase in agricultural output and creation of a demand for organic produce outside the kibbutz).[94] When I interviewed him, he explained it as follows:

> Let me explain to you what guided us. You need to understand that conventional agriculture is actually based on industrial production. It's an industry without moral boundaries, where everything is allowed. If you want to spray [chemical pesticides or fertilizers]—no problem, do whatever you want. Organic agriculture is also an industry, but it has morality and limitations. And regarding the biodynamic agriculture [that was practiced at Kibbutz Harduf]—well, this is agriculture which

has nothing to do with industry and markets. There are all sorts of do's and don'ts that have no realistic logic. I respect that. [Biodynamic agriculture] is good as an educational practice, but that's not the way to do serious farming. It might be appropriate for some remote and marginal farms in Germany or France, but it does not suit Israel and the agricultural and food marketing systems here. [When we established HOFP] we turned it [agriculture in Harduf] into regulated and organized agriculture, one that is also suitable for industry.

Shapira's words are in line with those of American organic farmers that sociologist Michael Haedicke defined as "the professionals," namely those who emphasize the potential for market growth in organic agriculture.[95] Likewise, sociologist Brian Obach, who discusses the "market versus movement" tension in the US organic movement, suggested the term "the spreaders," to describe those who seek to expand ideas of organic and sustainable agriculture and who thus accept and advance the corporatization (and conventionalization) of the organic sector.[96] Therefore, Shapira's words might also fit the discourse of "the spreaders," to use Obach's terminology. However, from a local perspective, his words illustrate, first and foremost, the logic that has guided the "new Israeli middle classes," namely professionals and people who held key positions in Israel's economy and culture who have been seeking to navigate Israeli society in a capitalistic-rational-instrumental way, which is liberated, ostensibly, from any "non-realistic," "non-rational," or "non-practical" ideologies.[97]

It should be noted that despite what Shapira shared with me when I interviewed him, in several press interviews he mentioned that his decision to get involved in organic agriculture was the consequence of a blood test in which a high concentration of toxins was discovered. He didn't mention the economic and commercial interests but instead described the establishment of HOFP as motivated from moral commitment, environmental care, and concern for public health. However, Shapira made sure that the health and environmental values of which he speaks were not framed as "alternative," so that his words could not be interpreted as "radical." He seemed to seek to avoid any aspect of "resistance" or "counterculture," and he rejected labeling HOFP as a firm that aims to undermine conventional food industries. In his description he demarcated clear cultural boundaries between HOFP and the other biodynamic farming and food production initiatives in Harduf:

> When I got to Harduf, I met amazing people, but they were completely dreamy. *Sroutim* [freaks or weirdos in Hebrew slang]. They were not realistic. They had tremendous ambitions, but their idealistic vision was *by far* [he said the words "by far" in English] high above them, they weren't "on the ground." They were imprisoned in their belief that there is no market for organic food in Israel. . . . Until we established HOFP, what was done in Harduf was considered simply as *hazui* [literally hallucinatory, a common word in spoken Hebrew used to describe something strange, bizarre, or weird].

Even after the establishment of HOFP, despite all of Shapira and his colleagues' efforts to fuse organic with the conventional, they were still "haunted" by images of weirdness and radicalism. For example, in 2006 the kashrut committee of Badatz HaEidah HaCharedit Jerusalem (a large kosher food certification organization) removed the kosher certification from the HOFP organic agricultural products, claiming that anthroposophy deals with "mystical meditation" that expresses a "Christian spirit" and "Christian cremations." "Anthroposophical Meditation," they said, "deals with attempts to connect with the dead in the spiritual world and beliefs of reincarnation."[98]

After the establishment of HOFP, and prior to the deal with Tnuva, Shapira's marketing strategy was to create an image of environmentalism, but not be associated with "tree huggers," or as he put it,

> We tried to educate the public to eat organic. We recruited environmental activists—serious guys, not just tree huggers—and a few students of alternative medicine, who became our "presenters" in conventional retail stores. They recited our mantras. We tried to explain to the consumers that they had to switch to organic nutrition. We tried to explain to them why organic food is so important, that eating an organic diet is much more nutritious than eating non-organic food, that it is important to consume organic milk produced in an organic dairy farm that benefits dairy cows and other examples and explanations of how organic agriculture is good for the environment.

But these marketing strategies didn't yield high revenues, and another change in Harduf's marketing concept occurred after Tnuva acquired the company in 2002.

## Organic Tnuva

As already noted, Tnuva, the food and agriculture enterprise controlling HOFP, is one of the largest food manufacturer/distributers in Israel. It was established in 1926 as a cooperative for marketing agricultural produce under nonprofit and egalitarian idealistic principles, according to which Tnuva should not make any profit and all revenues would be transferred entirely back to members of the Jewish settlement movement in Palestine.[99] During its first several decades, hundreds of moshavim (cooperative agricultural settlements) and kibbutzim joined in the cooperative ownership of Tnuva and received a share in the association. Farms transferred all their produce to Tnuva, and the kibbutzim and moshavim received payment after the produce was sold and a commission for Tnuva was deducted.[100] For many years, Tnuva was conceived as *the* ultimate "Israeli brand." The company branded itself accordingly, as exemplified in one of the company's slogans: "Tnuva—growing up in an Israeli home."[101] During the 1990s and the beginning of the 2000s, Tnuva expanded beyond its initial activities, which were limited to the distribution of fresh (conventional) agricultural produce (including dairy products and meat). At that time, the company acquired small food production companies, including companies that produced processed and frozen foods. This activity increased the company's production and sales volume and paved the way for its status as the largest food conglomerate in Israel.[102] The relationship between Kibbutz Harduf (then the representative of countercultural agriculture) and Tnuva (the epitome of the mainstream, conventional, and industrial food production in Israel) began in 1998 when the latter sought to market organic milk.

Prominent scholars of culture and society, such as Pierre Bourdieu, Dick Hebdige, and Thomas Frank,[103] maintain that encounters between the "alternative" and the "mainstream" are often characterized by the pressure exerted on the "alternative" (or what counts as "counterculture" or "heterodoxy") to be assimilated into the "mainstream." When "business culture" meets "counterculture," the latter often faces the dilemma of whether it should be rearranged in more heterogenic ways, namely changing what is conceived as alternative modes of production into hegemonic ones (and thus gaining access to a larger audience/more consumers), or in an autonomic way, which may tie the producers and their products to the margins of cultural production. A few Kibbutz Harduf members—Shapira among them—insisted that HOFP should move toward the mainstream and claimed that if Kibbutz Harduf did not accept Tnuva's suggestion of merging, its organic agriculture would go down the drain. After much debate, the final decision was made in accordance with Shapira's approach. In 2001, Kibbutz Harduf's food-related holdings were sold to Tnuva, including the rights to use the brand name "Harduf."

The agreement between HOFP and Tnuva specified that the parties would jointly manage the agricultural initiatives, which were based entirely on organic methods: orchards, field crops, and a bakery. HOFP also sold Tnuva its packing house, and the latter developed it into a department in the HOFP firm that specialized in packing and distributing imported organic grains, pastas, flours, and legumes. Since 2002, Tnuva has produced, marketed, and distributed a range of organic dairy products using the brand name Harduf. Tnuva's website claims, "Tnuva produces, markets and distributes a range of organic dairy products under the brand Harduf, which it produces at the Harduf dairy located at Kibbutz Harduf. The milk is received by the dairy from the only organic cowshed in Israel."[104] Up until 2012, when Tnuva decided to reduce its share in the activity of HOFP, Tnuva managed the distribution of all of Harduf's products. The offices of HOFP (renovated to be state-of-the-art after the sale to Tnuva) remained in Kibbutz Harduf, even though they were now the property of Tnuva. The dairy farm remained under the ownership and management of the kibbutz, but all the dairy products produced there have been sold to Tnuva, according to the agreements. Shapira described the changes that took place after the merger between HOFP and Tnuva as follows:

> We switched to a strategy based on the premise that whoever decided to eat organic came up with the idea on his [her/their] own, without us, and that he [she/they] got it from a journal article, or got to know what organic food is on a trip abroad, or maybe from a nutritionist who recommended it, but we, as a company, do not need to convince them that organic is better. We stopped the public activities and our efforts to explain and educate and focused on putting produce on the [supermarkets] shelfs. Anyone who wants organic—we will make sure [to provide him/her/them] a wide range of organic products.

When speaking about HOFP, Shapira used neoliberal terminology that emphasized consumer choice. In a newspaper interview he also testified to the relationship between organic agriculture and big conventional agribusinesses:

HOFP, under the management of Tnuva, increased organic consumer choice. The Israeli field of organic food has grown over the past three years at a rate of 25% a year [an increase in consumption], and if it were not for Tnuva it would be half of what it is today . . . This is a win-win situation. Tnuva benefits from the organic by learning how clean and healthy food production should work and by being introduced to alternative production techniques. And we [HOFP, the alternative food initiative that went through a process of conventionalization] can reach a greater number of consumers, who consume healthier food. We all benefit from it, even the environment.[105]

## Distribution and Consumption:
## Harduf and New Age Organic Food Consumers in Israel

Anthroposophy is one of the most popular New Age doctrines and lifestyles practiced in Israel. As part of the fieldwork conducted for the research on which this book is based, I interviewed 12 consumers who identified themselves as affiliated with various levels and aspects of anthroposophical theory and adopted them into their daily lives. In addition, while visiting sites where organic produce was sold and attending healthy food and ecologically themed events, I informally interviewed several other Israeli New Agers. Many of them testified that they consume organic food regularly and expressed a preference for organic products branded with the Harduf label. For example, Dorit said,

> We buy organic packaged food at Rami Levi [a conventional retail store chain]. There are all sorts of organic products there, but we prefer those of Harduf. [When I buy Harduf products] it's a bit like connecting myself to the kibbutz.

Most of my interlocutors espousing anthroposophy—whether they were residents of Kibbutz Harduf or followers of anthroposophy who lived in other places—identified the brand name Harduf with Kibbutz Harduf, not with Tnuva. Of those I interviewed who were aware of Tnuva's role in Harduf products, none were critical of the collaboration. Even when explicitly asked about it, they said that they thought this collaboration was reasonable and acceptable. Narkis, who specializes in anthroposophical education, even praised it:

I don't think that [HOFP being controlled by Tnuva] makes Harduf [HOFP] any less organic. . . . I think it's even better, because with the help of Tnuva the access to healthier food can be expanded. But the truth is that there are things in Israel that concern me a lot more. For example, education, which is, by the way, another issue that Harduf excels at.

Indeed, the main drive for the popularity of anthroposophy in Israel is education and the popularity of Waldorf (or Waldorf-inspired) schools and preschools.[106] In this educational system, a lot of attention is directed at the food served to the children, and the curriculum even includes extensive references to nutritional studies, agriculture, and cooking. Dalia, the appointed menu developer in one of the Waldorf education institutes, explained:

> We explain to the children what organic food is and how important it is for their health. . . . The food in our preschool is not one hundred percent organic, but it is vegetarian and includes only healthy foods. . . . The food is meticulous, healthy, nutritious, and vegetarian, strengthens, and nurtures. . . . True, it is not completely organic, but it is good and healthy food.

Dalia's remarks indicate that the Waldorf educational system serves as an intermediary translating the abstract concept "organic" into "healthy and vegetarian" for young children.

Moreover, the process of formulating the Israeli culinary-anthroposophical repertoire is based on "cross-cultural consumerism,"[107] or introducing individuals to ideas, experiences, and knowledge about "other" cultures. For example, Gila, a 49-year-old resident of Kibbutz Harduf, describes her first encounter with the connection between food, eating habits, and anthroposophy while abroad:

> I started studying anthroposophy in Berlin. In Israel, at that time, no one knew anything about it. Take the topic of ecology, for example. There [in Berlin] I learned about the importance of the subject. . . . Then I lived for five years in the San Francisco area. Everyone there talked about healthy food and stuff like that. . . . Today it has become completely mainstream, eating healthy and organic food. But I was already dealing with nutrition and ecology 30 years ago.

Ecology, as Gila stated in her remarks, is an important component of the anthroposophical philosophy. Anthroposophical ideas relating to agriculture as a holistic practice are part of a wide ecological discourse, which is interwoven with New Age culture. This discourse is based on the recognition that human beings have an obligation to care for nature and its conservation and to strive for a society that lives in harmony with nature.[108]

Thus, one would expect that organic food would be an example for realizing theoretical anthroposophical ideas and that progressive projects and alternative actions relating to food and the environment would be woven around the organic in areas where people who espouse this doctrine live. However, at the heart of the anthroposophical discourse lies the concept of "personal transformation," and socio-environmental change is subject to personal change.[109] The words of Sivan, a New Ager and organic food consumer aged 32, expose the predominance of the "personal" on environmental ecology and community:

> We just buy organic out of habit. . . . Listen, you ask about ecology and fair trade, and it all speaks to me, but it's not in my practical awareness. . . . What is important to me is that the food will be full of life, food that nourishes you both spiritually and physically. . . . I truly believe that everything begins in our body.

The expression "food full of life," as well as similar expressions such as "food that heals the body," was repeated by many other New Age consumers. The main meaning of organic food—as arises from their sayings—centers on individualistic spiritual and physical nourishment. Moreover, the act of consuming organic food is carried out on the personal level and not integrated into collective action. A description of organic food as a means of connecting the personal body and soul, and as a source of "positive energy" and personal "energetic nourishment," was repeated by many of my interviewees who espoused anthroposophy. These statements are consistent with perceptions and dogmas accepted in New Age culture.

The British sociologist Steve Bruce claims that New Age culture is characterized by an alliance between individualism, self-sanctification, and care for the "other" (referring to "the other" in its broad sense, including care for nature and environment).[110] To use the terminology suggested by Bruce, the significance given to organic food as healthy illustrates the centrality of individualism and self-sanctification in the fundamental levels of the consumption practices of Israeli New Agers. Concerns for the "other" remain, mainly, at the symbolic level. In other words, social-environmental values are often declared by adherents of Kibbutz Harduf's commodities, but in practice, these values are marginalized and interwoven with the New Age imperative of self-care.

To conclude, food-related initiatives in Kibbutz Harduf—which were originally entrenched with egalitarian and anthroposophical ideologies and intended to serve as a source of livelihood for the members of the kibbutz—have changed over the years. Nowadays, they seem to represent a division between the two aforementioned approaches regarding organic food production and distribution: the "organic mirror" approach and the "alternative" approach. On the structural level, the different modes of organic food production and distribution in Kibbutz Harduf seem to be a clear example of the so-called "bifurcation of organic," referring to "the distinctions between large growers and corporate organic distributers—those who use industrial and conventional practices and aim at the mainstream market, and small growers, tending toward more agro-ecological methods."[111] Looking, however, at the cultural level, and focusing on the logic that has been guiding organic food distributers, producers, and consumers in Harduf, reveals that there is much more in common than distinctive elements between them. Harstein, Shapira, and the New Age consumers related to Harduf all seem to be constantly seeking to cross the boundaries of the kibbutz. As such, they all use "organic" as a means of integration in various social fields that operate outside the kibbutz's boundaries.

As described above, HOFP, Harduf Restaurant, and Eden Teva Market operate as actors who aspire to reach reconciliation with the mainstream Israeli food sector and even to position their organic produce at the heart of the broader food-related fields. Furthermore, in considering the meanings ascribed to the notion organic by the consumers of these distributors, it seems that the organic food initiatives in both spheres of distribution—the "kibbutz" and the "city"—have contributed not only to the corporatization and conventionalization of organic food in Israel but also to the promotion of neoliberal-individualistic lifestyles that are characterized by the emphasis on consumer choice, self-health, and self-care. The following chapter will show that an extension of the meanings ascribed to the notion of organic occur when it is discussed in other social fields: the media and the political field.

Chapter 5

# Translating Organic

## Cultural and Institutional Intermediaries and the Production of a Post-National Organic

"Can someone explain what organic agriculture is exactly? Could anyone, for God's sake, give a precise definition of organic?"

—Herbert Zinger, an official of the Ministry of Health, in a meeting of the Economic Affairs Committee of the Knesset, the Israeli Parliament, that dealt with a bill on the regulation of organic agriculture, June 2000

While the previous chapters dealt with spheres of production, consumption, and distribution, this chapter shifts our attention to the ways in which the symbolic meanings of the term *organic* are formed beyond these spheres. It focuses on two fields that appear not necessarily to be related to organic food and farming, or even related to each other: media and law. However, a closer look at actors in these fields reveals their important impact on the field of organic food as well as the common denominator of their discourses and practices. In both of these fields—media and law—actors work to (1) mediate between producers, distributers, and consumers and (2) "translate" the term "organic" so that it can be assimilated into the local context.

In this chapter, I describe the ways in which the notion of organic has been culturally and institutionally translated and mediated in Israel. I focus on individuals who have worked to create connections between the levels of production, distribution, and consumption to ask what socio-cultural trends and ideologies they have promoted. What are the cultural frames and repertoires[1] they produce? How do cultural actors—those who are referred to as "cultural intermediaries" (or "arbiter elegantium" to use the terminology suggested by Pierre Bourdieu[2])—design the portrayal of "organic" in Israeli mass media? How have cultural intermediaries dealing with organic food constructed environmental perceptions, encouraged the adoption of eco-habitus, and cultivated a "lifestyle of health and sustainability (LOHAS)"?[3] How do institutional intermediaries understand the notion of "organic" and implement their understanding through regulatory policies related to organic food production in Israel? In order to approach these questions, I return to the notion of social (cultural and institutional) intermediaries as well as to the notion of translation, the displacement and reconstruction of knowledge, ideas, and similes.

## Cultural and Institutional Intermediaries

Cultural intermediaries are the tastemakers who define what counts as good taste in today's marketplace. As Julian Matthews and Jennifer Smith Maguire explain, "cultural intermediaries perform critical operations in the production and promotion of consumption, constructing legitimacy and adding value through the qualifications of goods."[4] They are individuals who are qualified in media, digital technology, and knowledge produc-tion. Acting as "tastemakers," they use their technological proficiency to influence others' perceptions and attachments. According to Bourdieu's account, "cultural intermediaries work on and through notions of cultural legitimacy, drawing upon their own cultural capital and dispositions in order to shape consumers' perceptions of and preferences for goods, prac-tices, and styles of life."[5] Sociologists Scott Lash and John Urry associate cultural intermediaries with what they term the "service class," whose members are preoccupied with issues of identity, performance, lifestyle, and openness to new experiences.[6] These are visual, auditory, and digi-tal media professionals, journalists, cultural critics, public relations and marketing professionals, and professional consultants in various subjects, including nutrition and health.[7] Callon and his colleagues suggest includ-ing in this category of "cultural intermediaries" all those who intervene in how consumers perceive and engage with goods.[8]

In addition to the aforementioned cultural intermediaries, I fur-ther suggest including lawmakers, representatives of state agencies, and representatives of certification agencies to the category of tastemakers, especially when we discuss foods labeled with ethical designations such as "organic." Like cultural intermediaries, individuals in these professions valorize and qualify goods and services and operate between the levels of production, distribution, and consumption. In the current era, which is characterized by the intensification of the flow of things, knowledge, information, and images,[9] the centrality of both cultural and institutional intermediaries is increasing and is reflected in the innumerable quantities of knowledge and information as well as "culturally translated" products, new goods, and new markets.[10]

Previous sociological scholarship on globalization pointed to the ways in which global ideas, practices, and goods are enacted and institutionalized into local contexts. This work describes how intermediaries are taking part in what sociologist Michal Frenkel termed "the politics of translation." The latter refers to a transnational flow of cultural objects and the dissemination of ideas and discourses beyond national and cultural borders. These processes of cultural diffusion shape, or at least have the potential to shape, the social realities in the societies in which new ideas and discourses become dominant.[11] Sociologists Michel Callon and Bruno Latour describe the process of translation as a formation of alliances between actors operating in different domains and striving to promote the adoption of a given model according to local and particular contexts.[12] As described below, it is both those traditionally defined as cultural intermediaries (journalists, professional consultants, and the like) and institutional actors (such as politicians and lawmakers) who together participate in the process (or politics) of translation.

The next section of this chapter explores how the notion "organic" is translated by producers of symbolic goods in contemporary Israeli media. The subsequent section deals with the mediation process that has taken place on the political field, focusing in particular on the Israeli Law for the Regulation of Organic Produce (which came into effect in 2008). It shows how states and institutional agencies played a central role in translating global ideas related to organic agriculture and promoted their assimilation along the lines of the Israeli (and the Israeli-Palestinian) political and cultural context. Together, these following sections describe the ways in which intermediaries in both mass media and the political field of law delineate symbolic and structural boundaries of, and within, the field of organic food in Israel. At the cultural level, the symbolic demarcation is based on (1) cultural cosmopolitanism and political post-Zionism, (2) the high status ascribed to urban lifestyles,

(3) individualism and self-care, and (4) consumerism and the burgeoning culinary culture in Israel. Applying a structural perspective reveals that both the political-economic ideology of neoliberalism and ideologies related to ethno-nationalism have informed the discourse and practices of actors in the media and the law as they engage in shaping the field of organic food in Israel.

## Mediating Organic

The four strands that comprise the coverage of organic food in Israeli media—cultural and political post-Zionism, an urban lifestyle, self-care, and hedonistic consumerism—can be exemplified by focusing on five media personalities to whom I will now turn: Dalik Wolinitz, Shiri Katz, Aviv Lavie, Rachel Talshir, and Ronit Vered.

### CELEBRITIZATION AND CONVENTIONALIZATION OF ORGANIC

Dalik Wolinitz is a well-known theater and film actor, television program host, and radio broadcaster in Israel. Following a heart attack in 1994 at the age of 37, Wolinitz began to express an interest in health food. He changed his personal eating habits, switched to organic food, and came to attribute his recovery and good health to these behavioral changes. Concomitantly, Wolinitz was exposed to the ecological meanings attributed to organic agriculture. Soon, he began to use his communication skills and access to Israeli mass media to vigorously promote organic food consumption and environmental care.

Wolinitz is probably one of the first Israeli "celebritus politicus," to use the notion suggested by Michael Goodman that refers to celebrities who attach themselves to problems of the environment and human development.[13] Dubbed "the number-one celebrity of the Israeli green community,"[14] Wolinitz developed an Israeli version of the Western "enviro-tainment genre" in mass media and promoted the trend of "celebritization of environmental concerns" in Israel.[15] Wolinitz is quite convincing regarding his sincere concern and motivation to solve environmental issues. As such, he has contributed extensively to the promotion of media discourses on environmental sustainability, risk, and health in Israel. "I've realized that what is going on here (in Israel) is actually genocide," he said in one of his press interviews. "They simply put dangerous toxins into our foods."[16]

In 1997, Wolinitz founded the Organic Food Consumers Association (OFCA), which set as its goal "safeguarding the interests of those who consume organic produce of any kind, by way of [voluntary] inspection of the quality and reliability of the produce, by maintaining a reasonable price level or by any other means," as he put it. When I interviewed him and asked about his activities in the OFCA, Wolinitz said, "As a representative and a spokesperson of the association, I publicly emphasize that I believe that those who switch to an organic diet not only become healthier but also promote, one meal at a time, significant environmental changes."

As the founder of OFCA, Wolinitz worked tirelessly. He made connections and collaborated with the Israel Bio-Organic Agriculture Association (IBOAA). For almost a decade—from 1999 until 2008, when the Law for the Regulation of Organic Produce in Israel came into effect—he presented and directed the OFCA, which was the only institution in Israel that would accept public complaints regarding any problem in organic food production as well as general problems of food contamination. He also made connections between consumers and growers, recommended small-scale stores to consumers who asked for recommendations, co-founded a private organic certification and inspection agency, and later served as a member of the board of directors of this agency. Despite working so hard in these areas, he invested most of his energy and time in promoting organic agriculture and preaching his belief in the advantages of organic food consumption in Israeli media discourse.

Wolinitz's media appearances took on a repetitive pattern. Time and again he told his personal story, which he presented as evidence of the virtues of an "organic diet":

> I was released from the hospital, and after a while, I decided to start eating organic. I met a naturalist from Amirim [a moshav in northern Israel that was established in 1958 by a group of people who created a community based on a vegetarianism, veganism, and resistance to conventional agriculture]. Soon I began to eat only organic vegetables, fruits, and legumes. My life changed completely. This diet even helped in opening clogged arteries. I was convinced of the benefits of organic, so I initiated the Organic Food Consumers Association.[17]

Just like an artist who devotes her or his body to art, whose public persona is an integral part of her or his work, and who blurs the boundaries between life and art, Wolinitz used his body and the history of his personal health as an apparatus to mediate between organic agriculture and his audience. His lifestyle and organic eating habits were presented in the health columns of lifestyle magazines as a "remarkable success" and an "amazing makeover" (in line with the ubiquitous makeover culture in Western societies).[18] His media presence, in which his personal story was entwined with explanations about organic agriculture and its relation to broader environmental issues, created a metonymy between environmental issues and a healthy lifestyle—one in which environmental care was refined as organic food consumption. Furthermore, his image as a friendly person, his past as a host for programs for children and youth (which ostensibly attests to his concern for future generations), and his known affection for animals and involvement in environmental matters all came forward during the time he worked as a mediator and activist in the field of organic food.

When I interviewed Wolinitz, and analyzed his media appearances, I observed his frequent use of the notion "repair" (*tikun*), referring to both the modernist-reformative sense of this term (willingness to do something about environmental and nutritional damage) and to the spiritual sense of this notion (such as *tikun olam*—the concept in Judaism of "repair of the world," or the New Age concept of the connections between "repair of the self" and "repair of the world"). According to Wolinitz, organic food consumption is "the easiest way to solve all kinds of problems that stem from industrial agriculture, including the care for the farmers' well-being," as he told me. In addition, his frequent invocation of anxiety-producing imagery is highly noticeable; among such instances of this terminology is "genocide," as previously quoted.

In 1998, he published an article titled "The State of Israel—Towards the World's First Organic State" in the IBOAA bulletin (distributed to organic farmers and consumers through natural food and health stores). In this article he presented his vision of how to elevate the status of organic agriculture in Israel:

> I have a dream, and this dream can be easily realized. [In my dream,] Israel's government announces the transition of all agricultural practices to an organic system, and within five years Israel becomes the first organic state in the world. In short order, conventional agriculture is completely discontinued. All the farmers undergo comprehensive training in organic methods and gradually switch to organic methods. . . . Organic restaurants and health centers pop up throughout the country,

serving organic food to attract tourists and medical tourists
from all over the world. . . . The Palestinian Authority joins
the project and the Arab farmers return to their farming traditions, to the organic farming of their ancestors, supplemented
by the experience and scientific knowledge of modern Israeli
[Jewish] agriculture and organic agriculture. Israel's economy is
thriving, and Israel gains recognition and support from all over
the world. The economic prosperity and the global support
help to establish real peace between us and our neighbors.[19]

Wolinitz's "dream" presents a vision that brings together national rhetoric and global-capitalist practices. It is a vision that stems from the
political ideology that has prevailed among Israeli middle and upper
middle classes—the social groups that have been the beneficiaries of
globalization. Sociologist Uri Ram argues that during the 1990s, global
capitalism, driven by the middle class and the economic elite, was the
driving force behind the Israeli peace policy (as reflected in the Oslo
Accords).[20] Wolinitz's utopian vision of an interlacing of organic agriculture, economic prosperity, and stable political security is compatible
with this post-Zionist discourse and ideology.

However, in reality, out of these two imaginary levels—the economic
level and the political level—Wolinitz's actions were focused solely on
the economic level. As time went on, the liberal-political images of
Jewish-Arab peace and collaboration through organic food production
decreased in Wolinitz's media discourse. Instead, he gradually came to
focus on mediating organic food through neoliberal terms. For instance,
less than a year after the publication of the above article, he published a
continuation. This time, he announced the beginning of "the realization
of the dream." However, he based his claim on examples that indicate
the conventionalization and commodification of organic as well as on a
growing depoliticization (in the context of the Israel-Palestinian conflict)
of organic agriculture. For example, he praised, in the subsequent article,
the acquisition of Harduf by Tnuva (see Chapter 4) and wrote,

[The deal between Harduf and Tnuva] is an expression of the
beginning of the realization of the dream about an organic
state. . . . Organic food is displayed across the country on
billboards. . . . OFCA and I contributed to this marriage.
We played an important role in the negotiations, and convinced Tnuva that the future is in "organic." We collected

and presented data from Europe and it convinced them. Big
retail chains are now interested [in organic food] and offer
more and more organic food, which is beginning to be sold
on a large scale. We are on the right path![21]

Thus, any indication of the realization of the political levels of his
"organic dream," or a statement expressing a yearning for it, are absent
from this later article.

As a reporter with strong connections in the Israeli food industry
and as the founder of OFCA—and someone also closely connected to
organic farmers—Wolinitz served as an active mediator in the early
stages of the acquisition of Harduf by Tnuva. The Tnuva-Harduf deal
had far-reaching implications on the field of organic food in Israel. It
served as a model for other acquisitions of organic producers and started
the process of industrialization, large-scale distribution, and consolidation
of organic food production in Israel. Wolinitz and his colleagues from
OFCA were also involved in those other deals. For example, in 2006, he
served as a mediator of another big acquisition related to organic food.
This time, Adama—then a small company engaged in importing, packing,
and distributing organic foods—was purchased by Maabarot Products,
one of Israel's biggest manufacturers and marketers of a wide range of
nutrition and health products. The irony is that Maabarot Products is a
hyper-capitalist company that was originally founded in a kibbutz—the
Israeli socio-agricultural institution that historically promoted socialist
and anti-capitalist values (see Chapter 4).[22]

Consequently, Harduf (controlled by Tnuva) and Adama (controlled
by Maabarot) became the leading brands in the conventionalized field
of organic food in Israel. Following the expansion of the demand for
organic food products, other companies that produce, pack, and market organic food products (such as HaSade Organic Products and Liv)
increased their organic production and marketing as well as the amount
of organic foods that they imported to Israel. This increase in the supply
of organic food products accelerated the establishment of organic retail
chains such as Eden Teva Market (see Chapter 4).

Thus, it is fair to say that Wolinitz, who strove to realize his vision
of an organic state, mostly contributed, in practice, to the conventionalization of organic in Israel. It should be noted that this realization of
"Wolinitz's dream" turned out to be the realization of a nightmare for
some of the actors in the Israeli field of organic food: the small farmers
and individuals who ran small shops adjacent to their farms. The latter

could not compete with the new corporate-organic companies or with the organic retail chains. Consequently, many of them went bankrupt. In an interview I conducted with Wolinitz, he told me that he is now aware of this matter, and thinking retrospectively, he rues his involvement in the commodification and conventionalization of organic.

Ultimately, throughout the years in which he was involved in the field of organic food, Wolinitz mediated the notion of organic in two critical ways: first, by encouraging individual organic food consumption as a practice of *tikun*, which can be seen as the Israeli vernacular of the neoliberal ethos of ethical-environmental care through personal-level consumption (consistent with the logic of "eating for change"),[23] and second, by harnessing his communication skills, social capital, and ability to navigate the public sector to pave the way for the commodification and corporatization of organic food in Israel.

## ORGANIC FAMILY

Wolinitz was not the only cultural mediator who used a personal story to promote organic food culture in Israel. Shiri Katz and Aviv Lavie, a couple in their late 40s, succeeded Wolinitz in this role. Since the second half of the 2000s, they have made a significant contribution to the presence of organic food in Israeli media. One of the communication techniques they have often used is to discuss their own family's eating habits and the place of organic food in their own kitchen and everyday life.

Katz and Lavie started their careers as journalists in the local newspaper sector (*mekomonim*). Media scholar Oren Soffer characterizes the mekomonim as media outlets that combine reports on "trivialities," references to nationwide and global events, and advertisements. These local newspapers assign plenty of space to personal columns.[24] Inspired by foreign newspapers, and especially by the "new journalism" style that prevailed in popular American magazines in the 1960s and the 1970s,[25] the mekomonim used natural and fluent language, reflecting a cultural environment looking to forge a new (postmodern) identity. Ultimately, the popularity of the local newspapers reflected the decline of the cultural-national hegemony in the field of journalism in Israel and the strengthening of Western and global cultural influences.[26] It should also be noted that these local newspapers made an important contribution to various fields of popular culture in Israel,[27] including gourmet culinary culture. For example, these outlets contributed significantly to the promotion of Tel Aviv as a culinary cultural center.[28]

Shiri Katz was first introduced to organic food while reviewing culinary trends for one of these local newspapers. She told me that she remembers the first time she heard about the notion "organic." It was while covering the culinary trends in Tel Aviv related to gourmet and imported cheese consumption. Interestingly, the first organic food she tasted was not a local organic food but an imported organic cheese. A few years later, Katz found herself engaged in importing organic food. She didn't deal with the import of material organic goods, however. As a journalist—a cultural intermediator—she dealt with the import of cultural ideas, specifically the meanings of "organic" as a concept and a culinary category. For more than a year, the *Haaretz*[29] magazine supplement *Galleria* (gallery), which covers issues related to fashion, leisure, popular culture, and art, published a weekly column on organic food written by Katz and her partner, journalist Aviv Lavie. The weekly column was titled "The Organics" ("Haorganim"). Later, the column articles were published as a guidebook: *The Israeli Guide to Organic Food* (2007).[30]

The weekly column and the book focus on the experiences of Katz, Lavie, and their children: a Tel Avivian family who began to consume organic food on a regular basis. The column also includes recommendations for places to buy organic food, contact information for organic growers, recipes, and a review on the quality organic foods. It also deals with a few of the key questions that concern many Israeli organic food consumers: "Why is it so expensive?" "How can we make sure that it is indeed organic?" "Will it make me healthier?" and "Is it tastier than conventional food?"

The image that emerges from their articles and book is that of an upper-middle-class Israeli couple. Their depicted lifestyle is consistent with what Oz Almog termed as a lifestyle of "post-Zionist yuppies," namely the creative stratum of Israeli society: educated individuals who were born in Israel to Ashkenazi parents. According to Almog, post-Zionist yuppies are characterized by a worldview that consists of an array of liberal beliefs. These include

> equal civil rights, competitive achievement orientation, economic efficiency, private initiative, career development and professional fulfillment, financial success, sensitivity to animals and the quality of the environment . . . freedom of thought and creativity, skepticism and criticism, self-awareness, emotional openness, feminism, intimate couple relationships . . . command

of information through the latest communications networks, aesthetics and health awareness, a refined taste, knowing what is going on in the world, [and] Western fashion trends.[31]

In their writings, Katz and Lavie share with their readers the ways in which their sociocultural-environmental worldview is translated into mundane practices and advise their readers to adopt those practices.

As their writings indicate, parenthood is seen as one of their major and most important engagements in everyday life. Among the values attributed to "post-Zionist yuppies" by Almog is a "protective but liberal parenting, a preoccupation with fulfilling family values and prioritizing the care for the children, often to the extent of developing a 'cult of parenthood.' "[32] As such, the task of raising children becomes the focal point of everyday family activities and a social practice of significant importance. According to Almog, this concentration on the nuclear family fuels the cultural individualism that characterizes this group and accompanies the decline of collective values. From a broader political-economic perspective, this phenomenon can be seen, as explained by sociologist Viviana Zelizer, as part of the Western "sacralization" of the child from a figure of economic worth to one of emotional pricelessness.[33] Previous research points out that normative expectations of parenthood in a contemporary neoliberal era—especially motherhood—include taking responsibility for being the moral and physical guardians of the next generation as well as fostering and promoting the child's potential.[34] As Cairns and her colleagues describe, the neoliberal discourses of childhood operate ideologically to enlist mothers in an endless array of strategies that strive for control over childhood. One of the primary means through which mothers strive to achieve such control is through consumption—the strategy most readily available amid neoliberal discourses of choice and individual responsibility.[35]

Accordingly, in their book and articles, Katz and Lavie describe at length their eating habits, which are based on organic food consumption, and the ways in which they have instilled healthy eating habits in their children. For example, Lavie described his first visit to the Orbanic Market:[36]

When I rolled into the house with my shopping cart buckling under the burden, the first thing I did was to organize a colorful exhibition on the table: there were Ziv Schwarzman's wonderful organic papayas of from Neta'im [a moshav in central Israel], radically sweet mangoes, a mountain of lychees, a hill of passion fruits, and a box of sabra fruits. For a few seconds the kids followed me silently with a puzzled look in their eyes, as if I had spread the latest version of Lego robots. Immediately thereafter, a fierce battle broke out between them over the "distribution of the spoils."[37]

In an interview I conducted with Katz, she said,

I think that my job [as a journalist] is to provide *nesting* [she used the word "nesting" in English]. We live in a world where we are all exposed to a lot of global knowledge. . . . Today, parents have no time to deal with the overwhelming information, to understand all the interests, to understand all the shticks that they are fed by all sorts of people with interests. My job is to "nest" them, to protect them from irrelevant information and to provide them with accurate information about what is really happening in the food industry.

Indeed, the main theme in Katz and Lavie's columns can be described as the connection between organic food and the (Tel Avivian and upper-middle-class) family. Their texts are accompanied by myriad visual images of "the normal family" (which is, according to these depictions, a heterosexual urban family) as well as of "family leisure" and "protective parenthood." These texts contributed to the association of organic food with semiotic meanings of "the healthy organic family." Katz and Lavie describe organic foods and dishes that are often liked by children, such as organic couscous, organic egg omelets, and organic ketchup, and present these foods as available "to the average Israeli consumer, not only to the fanatics of organic," as they put it. Thus, they seek to associate organic food consumption with "normal" familial practices. In their words:

At the entrance to our apartment, there is no doorman who makes a selection and throws out every food item that does not have an organic label on it. Far from it. We read with great interest [previous newspaper articles dealing with organic food], we agreed with quite a few things and didn't agree with others, but mainly we found it difficult to connect to strictness. We are not Orthodox.[38]

For Katz and Lavie, the recipient of their messages is the private consumer, particularly the urban, educated consumer, who knows something about ecology, sustainability, and healthy diet, the one who struggles with a busy schedule and ambiguity regarding mundane practices.

After the period in which they worked on the weekly column and after the publication of the *Israeli Guide to Organic Food*, Lavie decided to dedicate his journalistic career to environmental issues, and Katz continued "to bring the organic issue to the mainstream culture in Israel," as she told me. In her journalistic work she continued to emphasize the perceived health benefits of eating organic food.[39] Together they discussed what they saw as a strong connection between individual choice, lifestyle, household consumption practices, and environmental and health-related issues. Thus, their work promoted the message of "vote with your money" and of the win-win (neoliberal) ideology that empowered the reflexive individual-consumer. Consumers, according to this approach, enjoy freedom of choice and thus should be directed toward making more "right," "rational," and "ethical" choices for the benefit of their individual selves, society, and the environment.[40] Based on the conversations I had with organic food consumers, it seems that this message resonated with their audience, who appreciated the celebration of the promise of a green economy.[41]

Katz and Lavie have never been criticized for advancing these aspects of green consumption, namely consumer practices that embody the "roll-back" of state responsibilities for public welfare and environmental protection as well as the "roll-out" of market and civil society attempts to fill these state responsibilities.[42] Katz and Lavie did receive, obviously, some critical responses, but these focused more on their high-class lifestyle and thus accused them of a patronizing attitude, of social insensitivity, and even of being "missionaries of organic food who work in the service of organic food producers and marketers."[43] Interestingly, some of the critical arrows launched toward them carried a nationalistic critique. For example, Yitzhak, a veteran farmer who grows both organic and non-organic produce in Kidron (a moshav in central Israel) told me,

> I don't like these articles in *Haaretz*. They just scare the public. You know what kind of reactions it brings? People start to tell jokes about organic, they say, "Do you know the difference between organic and non-organic? That in regular [conventional] agriculture they [the farmers] spray [pesticides] in the daylight and in organic agriculture they spray at night."

I do not think it is necessary to frighten people; it will just keep them off the organic. . . . How can [Katz and Lavie] just cancel, in a flash, all the amazing achievements of Israeli agriculture? Listen, I am both organic and conventional [as a farmer], I know all sides and I can tell you—we have excellent agriculture in Israel, to the glory of the State of Israel.

In addition, Yitzhak condemned Katz and Lavie using an expression that reflects existing national-ideological tensions and fits Oz Almog's adjective "post-Zionist yuppies": "I think that they [Katz and Lavie] have forgotten what Zionism is!"

## Organic Nutritionism

The work done by Rachel Talshir follows several cultural patterns identified in the works of Wolinitz and Katz and Lavie. Since 2005, she has been publishing a weekly column entitled "To Eat or Not to Eat" in *Haaretz*. The column deals with issues related to "eating right."[44] In 2012, several of her articles were compiled into the book *The Secret Method: 100 Simple Rules for Healthy Eating*. Soon after its publication, the book became a bestseller.

When she was in her early 40s, as she describes in her book, she was diagnosed with cancer. After she was treated and passed the recovery period, she became engaged in health-related issues and health food consumption. In a lifestyle magazine article where she was interviewed, the author noted, "At the age of 52, and after a documented struggle with cancer and a drastic change in her diet, Talshir looks perfect: shiny skin, a bright smile, and an excellent figure." The hallmark of Talshir's column and her book are sketches/caricatures of herself, which try to depict her as younger-than-her-age with a slim body and a white-European look. It is noticeable that Talshir—like Wolinitz—recruits her personal story and uses her personal self as an authentic testimony to her arguments regarding "what to eat" and proof that adopting her eating habits promises a youthful appearance and good health.

In her book and in her newspaper articles, Talshir aims to explore "the nutritional revolution, which began in Western countries at the turn of the third millennium."[45] According to Talshir, the "nutritional revolution" includes "four dietary methods, which are popular nowadays among many around world."[46] These methods are endowed with multiple virtues and a promise that those who adopt them will find that "it is easy" to stop eating junk and take care of their health, but at the same time, they will not lose their "joy of life and sense of meaning," as described on the back cover of her book. Talshir portrays these diets, or "eating methods," as a complex doctrine that includes "do's and don'ts," explanations about recommended eating habits, and an in-depth description of the health benefits of the recommended food products as well as the ills they are supposed to cure. Among these methods ("*shitot*," as Talshir put it, using the Hebrew word for "methods"), which are known and practiced by many health food proponents in Western countries, are the Kingston Method (Shitat Kingston),[47] the Wigmore Method (Shitat Wigmore),[48] and the Boutenko Method (Shitat Boutenko).[49]

One of the methods she discusses is called the Conscience Method, or the Moral Method (Hashita Hamatzpunit). This method, according to Talshir, stems from three movements: "Those who fight to protect our planet, those who fight for justice, and the animal rights movement." She explains, "At the basis of this method lies the idea that one cannot do damage without being damaged, and vice versa—those who protect our earth, care for labor rights and for animal rights are also those who benefit themselves and their health." One might have expected that organic food, a category often characterized by the principles of health, care, ecology, and fairness, would be significantly represented on the pages of the book where Talshir elaborates on the Conscience Method. Instead, organic food is mentioned in relation to other nutritional methods, which focus on extensive consumption of fruits and vegetables. For example, in the discussion of a plant-based diet, Talshir refers to organic food in this way: "The question of whether organic food has or does not have nutritional advantages over its regular equivalents [conventional foods] has preoccupied the food industry in recent decades. . . . According to the Wigmore and the Boutenko methods, the quality of the soil has a tremendous impact on the health of those who eat the plants that grow in and on it."[50] For Talshir, based on her understanding of the Wigmore and the Boutenko Methods, organic plants are the only foods that grow in a non-contaminated and protected soil, and therefore they are particularly healthy and nutritious.

This emphasis on the nutritional virtues of organic food is anchored in a perception of foods as a resource for nutritional elements and of eating as an act of consuming nutritional ingredients. Talshir's approach to organic food, as well as to food and eating in general, is in line with what sociologist Gyorgy Scrinis calls "nutritionism"—a dominant approach among nutrition scientists, dietitians, and public health authorities that gradually reached the general public. Nutritionism refers to a reductive understanding of nutrients as the key indicators of healthy food, an understanding that has been developed at the expense of other levels and ways of engaging with food.[51] Scrinis points out that "the nutrition industry has implicitly or explicitly encouraged us to think about foods in terms of their nutrient composition, to make the con-nection between particular nutrients and bodily health, and to construct 'nutritionally balanced' diets on this basis."[52] Consequently, he argues, nutritionism has given rise to and promoted the nutritional view, the emergence of the "nutrition-conscious individual," or the "nutricentric person"—that is, the person who strictly follows nutritional "require-ments" and sees food as something that must be measured, monitored, and scientifically managed. The "nutricentric person" is also the one who feels empowered to understand the nutri-biochemical basis of food and the body, to follow dietary debates in the media, and to see through the more inflated nutritional marketing hype in order to make informed decisions on putting together a "healthy" diet. The "media frame" that Talshir offers in her "To Eat or Not to Eat" column[53] has contributed to the assimilation of this nutritionist approach as a core frame toward understanding of the notion organic (and of course, toward food and eating in general).

Despite her discussion of social issues in reference to the Conscience Method described above, Talshir rarely refers to the local conditions of food production or to political issues related to local food. Her focus on health-related issues is limited to a discussion on the health of the con-sumer and rarely addresses concerns related to, for example, the health conditions of the 22,000 Thai migrant farmworkers who do the bulk of the labor in Israel's agricultural sector, including the corporate-organic sector.[54] Structures of inequality in the Israeli food system are beyond her scope. Unlike Wolinitz, Talshir rarely deals with issues related to the macro-level of the Israeli food system, does not present any polit-ical vision for changing the food system, and most of the time does not refer to mundane issues beyond nutritional questions (such as the meanings of food and eating in relation to parenthood, as addressed by Katz and Lavie). Even issues related to environmental care are presented as merely a means for the production of healthy foods instead of as an end in themselves.

As implied from the subtitle of Talshir's book, *100 Simple Rules for Healthy Eating*, her cultural mediation seems to be contributing to the construction of an Israeli version of the global popular genre dubbed "how to eat."[55] Julie Guthman presents a thoughtful critique of this genre, which appears to have originated in the Global North at the beginning of the last decade. Guthman argues that this genre draws heavily on nutritional studies, creates (paradoxically) uncertainties and ambiguities regarding the question of "what to eat," neglects social problems in food systems, draws on Manichean ethics (industrial foods are conceived solely as "bad" and "natural food" solely as "good"), emphasizes neoliberal individual choice, and presents an anti-political approach that devolves regulatory responsibility to consumers via their dietary choices.[56] This critique seems relevant to Talshir's writings. In her articles, and in her book, the "good" is represented by plant-based foods. Organic plant-based products are presented as the crème de la crème in this respect, while the "bad" is represented primarily by sugar and its substitutes, sugary foods, animal products, and, of course, industrial and processed foods. In order to succeed in adopting "good" eating habits, and to be saved from the ills of the growing contemporary "bad" nutritional habits, the column's readers are asked to experience all recommended dietary methods, to acquire in-depth knowledge and nutritional understanding, to mix and match from the suggested *100 Rules for Healthy Eating*, and to practice new and healthier eating patterns that work for them. In one of her articles, Talshir argues, "There is no one you can trust. We have no choice but to accept responsibility for ourselves [as consumers] and to examine for ourselves everything that makes it to our plate and then goes into our mouth." She also says in her book,

> Every person must be familiar with [the healthy nutritional practices], experience them and choose from them according to what works for him, according to what makes him feel better, what suits his lifestyle and outlook. Think of a tourist who travels all over the world and collects memories and smells, and later when he returns to his home, he carries those that are dear to him [it is written in masculine form in the book] and saves them for himself, and adopts them, according to his taste, a little from every place he visited: experiences, customs, recipes, a few mementos and longing.[57]

Thus, Talshir's version of "how to eat" not only fosters concentration on self-care and on individualistic-consumer questions but is also based on openness toward global nutritional discourse and toward what is conceived as alternative, healthy diets in the global media sphere.

Organic Food and the Import of Foodie Culture to Israel

Despite the differences between them, the common denominator in the ways in which Wolinitz, Katz, Lavie, and Talshir address the issue of organic food in Israeli media is their efforts to portray the idea of "organic" in utopian-ideological terms. But organic food is also described in contemporary Israeli media in non-utopian, hedonistic contours. One clear example is an article titled "The Organic Cockcrow," which was published in the *Haaretz* weekend supplement *Galleria*. In the article, organic chickens grown in Israel are discussed as follows:

> In the West, the place where large-scale industrial agriculture and the supermarket culture developed, they preferred to stay blind to the ways in which the product got to the shelf. . . . In modernity, people are already used to the unnaturally monstrous sized fat chickens. . . . Organic chickens are smaller in size than those grown in conventional agricultural coops. The percentage of fat in their body is also lower, and accordingly the meat is a bit drier than we got used to, but that is the real wonderful taste of chicken meat. . . . Our goal is to trace back, and to get as close as possible to what nature intended to create for us. For this reason, and in order to obtain the desired quality of meat, the [organic] chickens feed on mixtures of organic grains imported from the United States.[58]

In this article, the author Ronit Vered collaborated with the celebrity chef Haim Cohen and the well-known cook Eli Landau. The latter added recipes for gourmet dishes to Vered's description of the organic chicken coop: Pollo alla Diabla (an Italian version of deviled chicken), Roast Chicken with Pomegranate Juice, Clay Pot Chicken Cooked with Garlic Cloves, and other chicken-based delicacies, which all included a strong recommendation for the use of organic chickens. "We love chicken meat," wrote the authors of the article, "we love it boiled, cooked, fried, or roasted. You can call it organic, but from our point of view, chickens that grow in good conditions, without unnecessary pressure and without growth hormones, simply taste better."[59]

Since 2007, Vered has published a weekly column, originally titled "The Next Pleasure" ("Haoneg Haba")—a name which suggests its hedonistic character. After a while, the title of the column changed to

"Dining Room" ("Pinat Ochel"). Her articles, which have been published not only in this column but also in numerous culinary magazines and books, deal with various topics related to gastronomy and often depict a travelogue of her experiences in culinary tours (for example, culinary tours in Acre and Jaffa, coverage of the culinary happenings in the Gaza Strip, a report from the Oxford Gastronomy Symposium, a dinner in the Galilee where roast pigeons were served, and an article about a new restaurant in New York that serves Jewish-Iraqi Kubbeh Soup). Ronit Vered represents a new generation of "culinary cultural intermediaries," namely restaurant critics, food journalists, authors of cookbooks, and, more recently, individual bloggers and writers of digital and social media.

The food journalism of Vered's predecessors, such as Amos Kenan, Ron Maiberg, or Shaul Evron, promoted conspicuous consumption of gourmet foods, directed audiences' attention to negotiating and resisting the hegemonic cultural imperative known as "Zionist asceticism," and legitimized "the good life" in Israeli culture.[60] However, Vered and her colleagues—the new Israeli (mostly Jewish) food writers—have adopted a more updated global model of culinary literature and journalism characterized by cultural omnivorousness. According to this model, the distinction between "high" and "low," and the boundaries between cultural repertoires, are blurred.[61] Currently, Vered is a prominent cultural intermediator who contributes to the construction of what is known as a "new Israeli cuisine."[62] She plays an important role in the Israeli gastronomic field, particularly in shaping a local version of global foodie discourse as well as the culinary taste of Israeli upper middle class (see Chapter 3).[63] Organic food is one of the many topics she deals with, and it is mentioned occasionally in her articles.

Although "foodies" is a category commonly assigned to individuals who are passionate about the pursuit of "good food,"[64] sociologists Josée Johnston and Shyon Baumann point out in their research on foodie culture in North America that contemporary Global North gourmet food writing stems from two interrelated patterns: (1) a manifestation of cultural onmivorousness within which "boundaries between legitimate and illegitimate culture are redrawn in new, complex ways that balance the need for distinction with competing ideology of democratic equality and cultural populism"[65] and (2) a predominance of political issues in contemporary food media that shape a different understanding of how food relates to equity, social justice, and sustainability.[66] These patterns are clearly evident in Vered's writings, which are replete with terms of political discourse, mostly in the sense of identity politics. She is inclined to cover artisanal food producers, especially those affiliated with marginalized groups in Israeli society.

Interestingly, she distinguishes herself from cultural intermediaries who champion food politics in the sense of health or environmental issues (such as Wolinitz, Talshir, or the American journalist Michael Pollan) by presenting herself as committed primarily to dealing with taste and pleasure. "Between Narcissus and Goldmund [referring to Hesse's famous novel that depicted the polarization between Narcissus's self-disciplined character and the passionate and zealous disposition of Goldmund], I belong to the hedonistic camp," she said in an interview I conducted with her. "I see myself as an agent of 'the good taste' and not necessarily for things related to health or ecology," she explained when distinguishing between her column and Talshir's or Katz and Lavie's. "Those whom I enjoy eating with and writing about are those who are also interested in organic food, but with a hedonistic objective. You know," she said jokingly, "these are people who love to eat and drink. These are not the sort of people who usually develop deep agendas. We consume organic vegetables and organic meat, but not because of any rigid ideology or strict worldview. Our approach to organic is from the level of taste."

Similarly, when she covered trends of CSA in Israel, she focused on the food itself and its culinary qualities. In the article "The Vegetable Boutique," Vered described her experiences as a CSA consumer:

> The weekly basket of vegetables we receive has turned into an inseparable part of our daily lives. . . . As food plays a central role in our lives, we appreciate the basket, and we think that it is of high value, but not necessarily because of its health or moral values. The vegetables that come to us every week are organic vegetables that are grown with excessive thought about humankind and the environment, but we will admit and not deny: . . . taste is the primary consideration among those who sanctify the delights of the palate. . . . In the pursuit of pleasures ["The Next Pleasure," as alluded to in the title of her column], we have sinned and harmed the earth and environment, and we will continue to sin. But when it comes to matters of the quality of the raw materials, our interests merge with those of the greatest knights of environmental care.[67]

In one of her articles, she covers the Orbanic Market and presents it from a distinctly gastronomic point of view:

> The life of a gourmand is difficult in the era of organic revolution. A wide range of considerations—nutritional, ecological, political, and social—pushed aside good old hedonism, and made old-school epicureans feel dazed and confused. One can think, "Is organic agriculture the solution to all ills of humanity?" Well, we do not have an answer to that. Cultivating an environmental conscience, instead of a beer belly, is a difficult task. All we can do is remain advocates of the good taste, to look from a distance, and to admire those who have succeeded in transforming their lives to a model of loyalty to earth and environment.[68]

Later in the article, she articulates the Orbanic Market as a hub of Israeli gourmet culture and associates it with popular celebrity chefs:

> Fresh zucchini, buttery avocados, bouquets of radishes, bundles of wild thyme, and other fresh vegetables and fruit fill the stand. These goodies will soon be on the tables of the guests of the new boutique hotel owned by Uri Jeremias [a celebrity chef, also known as Uri Buri] in Acre. Five years ago, when they planned to build a luxury boutique hotel, the Jeremias family decided to produce the raw materials for the hotel's kitchen themselves and established an organic farm for this purpose. It turned out that the farm provides much more than the needs of the hotel. Thus, they opened the stall in Tel Aviv's new organic market, which is one of the main places where one can get the Jeremiases' organic produce.[69]

In Vered's articles, Orbanic Market is not presented as a family-friendly site nor as a place that befitting household consumption (completely juxtaposed with the ways in which the market is described by Lavie) but as a culinary-touristic site[70]—"a good place to try new tastes and to find exotic foods," as she puts it.

In an interview, which took place about three years after the article about Orbanic was published, Vered told me that she gradually lost interest in that market and what it had to offer:

In terms of produce, it was less and less exciting. I didn't like their obsessive preoccupation with organic certification [of the growers and the market operators], with organic standards, is it organic? Is it not organic? I do not know. It just doesn't really speak to me. I find it so boring—the discussion about pesticides, and the types of fertilizers. I understand the importance of organic, but all the discussion about chemical, biological processes . . . it simply ruined the experience of visiting the market.

In the same interview she said,

I'm not comfortable with extremists, with those that wave the organic flag. Fanaticism doesn't work for me. [When farmers] work only to promote the organic agenda and not for good food—I'm out! It's not for me. Don't get me wrong, I think that we must encourage small farmers. . . . I adore so many things they do: the weekly letter they send, the way they grow the vegetables, the fresh vegetables covered with mud, the fact that they provide seasonal food, all of this—I just love it! It stimulates one's creativity. . . . For example, Yoav [a farmer from one of the wineries in the North] has an organic plot and we talk a lot about organic [food, agriculture]. He is a farmer who is interested in organic; he does it in a very sane and reasonable manner. He is, just like me, in search of the good taste.

Nevertheless, Vered based her idea of "good taste" on a detailed description of the people who make the food and on an attempt to expose the social reality in which they operate. Depictions of "food with a face," which is a common discursive strategy in Global North gourmet food writing,[71] is a writing strategy Vered uses repeatedly, including in the articles in which she discusses organic food. This technique for describing food is particularly demonstrated in her depictions of fruits and vegetables. The latter, ostensibly, lack any uniqueness or enchantment, but after she "personifies" them, the organic vegetables seem re-enchanted. For example, she writes,

Twelve years ago, Sa'ar Sela, a fourth-generation descendant of Kfar Tavor's farmers and vintners, began cultivating his

ancestral land. At first, he followed the conventional path
of farming, similar to his ancestors. Two and a half years ago,
he started to transform the soil to organic. The farm workers
gather around for breakfast. On the table one can find a fresh
pitcher of pink grapefruit juice, a bowl of sweet fresh pea
pods, picked recently from the field, cauliflower with goat
milk yogurt, celery salad, fennel salad . . . and other kinds
of salads, dips, and wild sourdough breads.[72]

And thus, organic farmers and other organic food producers are described
by Vered not only as cultivating organic food but as cultivating "good
taste." Vered therefore demarcates organic food as part of the burgeoning
culinary culture in Israel. Unlike Wolinitz, Katz, Lavie, and Talshir, she
deliberately states that she is less concerned about issues related to health
or environment (and thus contributes to processes of depoliticization of
organic food in Israel). Yet the common denominator between those five
cultural intermediaries is that they all overlook structural issues in favor
of consumer-individualistic aspects, such as nutrition and taste.

The modes of "translating organic," as presented by cultural intermediaries such as Wolinitz, Katz, Lavie, Talshir, and Vered, are also
evident in, and seem to have been affected by, the ways in which this
notion was interpreted and institutionalized in the political sphere. In
the following section, I reflect on endeavors for the establishment of
a state-controlled organic law in Israel. Quite similar to the cultural
frames of organic in popular media, the legislative discourse on this
issue minimizes the meanings of *organic*, formulates organic agriculture
into a niche sector of "nutritional-premium foods," and reinforces the
export-orientation of this sector, rather than viewing it in the frame of
broader ideas such as environmental concerns, public health, land use,
social justice, etc.

## Organic Law

The legislative codification and governmental regulation of organic
farming in other parts of the world is often conceived as the pivotal
moment in the process of institutionalizing organic farming movements.
Researchers who have analyzed the processes of standard-setting, labeling,
and formulating regulatory tools for organic food production describe
them as processes to rationalize the abstract philosophical environmental

ideas embodied in the notion organic.[73] Critical scholars argue that these processes have often subverted the original socio-ecological vision of sustainable, locally scaled, and equitable food provisioning, and have led to the "mainstreaming of organics."[74] Examination of the processes that led to the establishment of the Israeli Law for the Regulation of Organic Produce (which came into force in 2008) from an organizational-institutional perspective reveals similar ramifications. However, it is in the cultural sphere where we can see the glocal logic that directed the enactment of the law as well as its impact on markets and environment in Israel and in Palestine.

The set of standards that came into full effect in 2008 to regulate the commercial use of the word "organic" in Israel was part of the alliance between Israeli organic and conventional farmers, state officials, and institutional agencies. These actors "translated" global ideas related to the notion organic in ways that (1) produced "cultural linkages"[75] between the "translated" notion (organic) and the Israeli local-political context and (2) made the regulation of organic food compatible with the economic and geo-political interests of these allied entities.

## "We Need a National Organic Law!"

It is possible that the Knesset, the national legislature of Israel, would never have enacted a law dealing with a minor agricultural sector, such as the organic agricultural sector that operated in Israel in the 1980s and 1990s, if the so-called Gush Katif crisis had not erupted (see Chapter 2). The exposure of this instance of fraud raised concern not only for the future of organic export from Israel but also for the Israeli agribusiness in general. Rafael Eitan (also known as Raful), then Minister of Agriculture and Rural Development (MARD) and Minister of Environmental Protection, appointed a public committee whose objective was to examine the regulation of organic produce. The committee's main recommendation was to enact a law for the regulation of organic food production and distribution. However, organic agriculture was perceived as marginal in the legislative field, and the enactment of the law dragged on for several years.

At that time, the IBOAA realized that organic farming in Israel—at the time a completely export-oriented sector—was in danger. Dan, a veteran organic farmer who had been active in the IBOAA since its establishment, told me in an interview,

When the Gush Katif crisis broke out, we felt we were finished. We [members of the IBOAA] realized that a legislative process must be pushed forward. We realized that we had no choice but to lead the whole thing. We understood that it was urgent to institutionalize the organic sector in Israel if we wanted to stay relevant players in the international market. We [the IBOAA] had to lead a change together with the Ministry of Agriculture and put together a law. Otherwise we knew we would be finished. We needed a national organic law.

Dan, his colleagues from the IBOAA, and other advocates (including Mario Levi, the founder of the organic agricultural field in Israel, and Wolinitz, who led the OFCA) used their cultural and social capital, as well as their positions in the field of agriculture in Israel, to contact officials from the MARD and from the agricultural lobby in the Knesset.[76] Together, they formed a coalition that promoted the passage of an Israeli organic law.

In trying to gain legitimacy in the political field and make sure that organic agriculture was not seen to be threatening conventional agriculture, they minimized the use of radical-environmental terminology and emphasized, instead, a Zionist agricultural ethos. In one of the early meetings in the legislation process, Yoram Porat, a representative of the IBOAA, said, "organic agriculture is going to live in peace with the other agricultural sectors."[77] Avshalom Vilan, then a politician who served as a member of the agricultural lobby in the Knesset as well as being a Knesset member from Meretz (a left-wing political party usually emphasizing a two-state solution to the Israeli-Palestinian conflict) decided to support the IBOAA and help them in promoting the law's passage. He felt sorry for the organic farmers—the Jewish settlers from Gush Katif—who feared that they might lose their source of income. Thus, despite his political affiliation, which was supposed to be hostile to the idea of Jewish settlement in the occupied territories, he said in a meeting of the Economic Affairs Committee of the Knesset,

I know that there is something in it that is at odds with my political view. But after the Gush Katif incident I talked to a few organic farmers [from Gush Katif]. They grow organic lettuce and all kinds of vegetables. Putting this incident aside, they [the organic farmers in Gush Katif] are honest farmers. These are devoted people. I do not agree with their political beliefs, but it is impossible not to admire what they developed there. Such a modern and successful system, we simply can't let it fall. They are connected to the land and we have to help them.[78]

This symbolic framework of modern Zionist Jewish organic farmers was mentioned repeatedly. Often, the modern and developed characteristic of the Jewish organic sector was articulated, implicitly or explicitly, as distinct from the allegedly non-organic Palestinian farming system. For example, in the Economic Affairs Committee of the Knesset meeting that dealt with the organic agriculture bill, Ilan Eshel, then the CEO of the IBOAA, said,

> One of our problems in Israel is trust and reliability. In Europe, they have faith, they are civilized there. We need this law for several reasons. [The law] will allow us to deal with "hitchhikers," farmers and traders who present themselves as organic, even though they really are not. When passing through Wadi Ara [an area in northern Israel in the Haifa District populated mainly by Palestinians] . . . one can see a huge oil mill plant, very near Umm al-Fahm [a city where nearly all the population is Palestinian]. There is a sign that says "organic olive oil." How can the oil possibly be organic? Does anyone know what goes on there? Does this seem reliable to you? I talked [to the owners of the oil mill plant]. All I requested from them was to take off this sign, because it is not organic, by any means! Do you know their answer was? They nodded their heads and told me: "You can say whatever you want, there is no law against this sign."[79]

The draft of an organic bill was submitted by Knesset members Shalom Simhon (from One Israel, then a center-left political party), Avshalom Vilan (Meretz), and Zvi Hendel (National Religious Party and National Union, both far-right-wing political parties) in 2000. When the Knesset discussed the proposed bill, Simhon announced,

> We hereby request to promote a regulation for organic produce in order to prevent disturbing incidents. We do not want the

Gush Katif incident to happen again, God forbid! We do not want the Ministry of Agriculture to hold another public inquiry for another issue in organic export from Israel![80]

It should be noted that external economic processes accelerated the parliamentary work on the bill. Before the Gush Katif incident and the beginning of the efforts to enact an Israeli organic law, the European Union presented to the international community the EEC Reg. 2092/91—a regulation on organic production of agricultural products and indications referring to labels on agricultural products and foodstuffs (EU-Eco-regulation). Furthermore, the IFOAM international accreditation program was launched. One of the criteria in the IFOAM program demanded a separation between organic growers' organizations and certification agencies. Gradually, the Plant Protection and Inspection Services (PPIS) of the MARD in Israel, which was responsible for supervising all produce exported from Israel—including organic produce—faced a demand for stricter inspection procedures from both EU authorities and IFOAM.

Consequently, the IBOAA was authorized by the MARD to establish Agrior (a private organic certification and inspection agency) as a subcontractor of PPIS for organic production regulation, certification, and labeling. This subcontracting process is similar to processes of economization of the peer-reviewed organic regulation procedure that occurred in other places in the world. Julie Guthman describes this process as "neoliberalizing labels, labeling neoliberalisms"—in other words, a creation of a market of organic certification where such a market previously did not exist, as a process of encouragement and enhancement of individualistic-voluntary sustainable practices.[81]

Back to the Israeli law: parallel to the pressure directed from Europe, the US federal organic law was developed, following the incorporation of the Organic Food Protection Act (OFPA) into the 1990 Farm Bill and the creation of a National Organic Program (NOP).[82] The NOP included a series of requirements and restrictions on the introduction of organic products to the American market. Yoram Porat, one of the representatives of IBOAA, discussed at length the US NOP in one of the early meetings of the legislation process:

I'll tell you what's going on in the world. The United States has decided that organic agriculture is a strategic sector that will push their agriculture forward into the twenty-first

century. This is what we should aim for. If we don't wake
up [and formulate a national organic law]—we will lose this
[the US] market.[83]

In the same meeting he also said,

Organic agriculture is strategically compatible with the
strengths and interests of Israel in the global market. We
should focus on *high-value products*, and organic products are
such [he said "high-value products" in English, signaling his
cosmopolitan and professional habitus] . . . therefore it should
be a national priority.[84]

It was emphasized declaratively, throughout the entire legislative process,
that the main purpose of the law was to provide PPIS with an effective
tool for better regulation of organic produce intended for export. Being
part of the global organic market, instead of, for example, building up the
organic market for environmental stewardship or care for human health,
was the main justification for the beginning of a parliamentary process to
establish this law. Avshalom Vilan framed it as part of a national project
of opening Israel to the world and expressed a post-Zionist[85] vision of
integration into global trade: "A country that has no [organic] national
standards and no national law will have a big problem competing in
the world market." On another occasion he declared, "When Israel will
have its own organic label, a label that will have legal validity, Israel
will join the enlightened countries of the world that produce organic
commodities and trade them in the global market." Similarly, the words
of the IBOAA CEO indicated that integration into the global circulation of organic produce was the prime consideration for the law: "The
importance of organic agriculture stems from several aspects. First and
foremost, the aspect of increasing revenues to Israel, since there is a
great potential for increasing exports."[86]

The centrality of export, as the main motivation for the legislative
process, is also reflected in the following dialogue. In a meeting of the
Economic Affairs Committee of the Knesset, Shlomo Kapua, the head
of the Agroecology Department of the Ministry of the Environmental
Protection, said, "We demand that the law will specify that we [repre-
sentors of the Ministry of Environmental Protection] will be permanent
members of a standards board [refereeing to an official advisory body
such as the US National Organic Standards Board (NOSB)]." The

chairperson of the meeting asked the participants, "Does anyone have any objection to this demand from the Ministry of Agriculture?" Tammi Mor, a representative of MARD, replied: "We have no objection, only that there is no such standards board." Doron Dinai, an attorney who represented the IBOAA, said,

> In Israel we have no need for such a standards board. The Israeli law adheres to international standards. Why? For one prosaic reason: most of the organic produce in Israel is exported. In any case, the countries that receive the produce demand that we comply with the accepted international standards. Even if we want to "reinvent the wheel" and set a unique standard for Israel, it will be as useful as demanding that Hebrew will be the official language for all the negotiations in the world.[87]

In addition to export, the Israeli lawmakers also facilitated the expansion of organic food imports into Israel. In one of the Economic Affairs Committee's discussions, which dealt with the design of organic regulations, one of the sections was read by a committee chairperson: "An organic symbol will be stamped on the package of each organic food product, indicating that the product labeled is an Israeli organic product." An official of MARD said, "We ask to omit the word 'Israeli' in order to include in the law imported organic products." This request, to omit the word "Israeli" and to open the (then) emerging Israeli organic field to imports, passed without reservation and without considering the inherent environmental and social flaw of importing food products.

One of the ramifications of the inclusion of imported foods in the Israeli organic law was the spread of organic supermarket chains in Israel. In an interview I conducted with a manager of one of the organic supermarket chains in Israel, I was told, "If we had not known that import would also be part of the organic law, my [organic supermarket] chain would not have opened. It was critical for us that [imported organic products] be regulated, with a legal organic label."

Other aspects that guided the design of the law were the dominant cultural frames of organic foods as associated with individual consumerism,[88] similar to the cultural frames produced by organic cultural intermediaries from the media sphere as presented above. Accordingly, organic agriculture was constructed as the production of luxury commodities, not as necessary for environmental preservation, public health, or

Figure 5.1. Organic supermarket in Israel. Photo by the author, November 2019.

the general prevention of dire consequences. The very few discussions dedicated to the impact of the law on the development of the organic sector in Israel were focused on the integration of what most of the participants considered a niche-consumer sector into the broader existing food system. In one of the Economic Affairs Committee's meetings that dealt with the bill, Avraham Poraz, the committee chairperson, opened the meeting with a provocative statement about the destructive implications of contemporary conventional agriculture and stated that the discussion about the bill was an opportunity to discuss and propose solutions to many problems in Israeli agriculture. He questioned whether organic agriculture would be a "real alternative" and seemed to expect to hear that organic agriculture offers solutions to all of these concerns. Eldad Landes, a representative of MARD, claimed that this question was irrelevant and insisted on talking about creating possibilities for consumer choice. Here's specifically how the conversation took place:

> [Landes:] Every person can decide for himself whether [he/she] wants to buy organic food or not. The Ministry of Agriculture does not want to decide which agriculture is better [organic

or conventional]. We want to promote the law so that it will do one thing: regulate. . . .

[Poraz, turning to him:] So, you're probably happy that we grow organic food in Israel and export 80% of it.

[Landes:] I am very pleased with that, and we want to support them [the Israeli organic farmers].

[Poraz:] But tell us—do you think that organic food is better?

[Landes:] We are not talking here about my beliefs . . . let me put it another way. Do you eat kosher food?

[Poraz:] In your understanding, the whole matter of organic is like kosher? Let me get this straight: you are saying that the Israelis and Europeans who buy organic food are actually suckers [*freiers* in spoken Hebrew]![89] . . . You and the Ministry of Agriculture do not believe [in the idea of organic] at all!

This vignette is emblematic of the sociocultural translation of the notion "organic" in relation to the Israeli context and the formation of an Israeli organic law. In this exchange of words, and in the following meetings of the committee, Poraz constantly tried to discuss the substantive advantages of organic agriculture over conventional agriculture and its ethical meanings. However, he did not succeed, and the discussions continued focusing on technical issues, such as creating mechanisms to protect global and local organic consumers who are exposed, allegedly, to frauds.

The concept of organic as expressed by Landes is very far removed from the ethical-ideological concepts raised by Poraz. The purpose of the law, as Landes noted, and as the law was eventually formulated, was to create a system of regulation with the purpose of generating trust in the label "organic" among consumers. According to his perspective (which is compatible with the "green economy perspective" presented by the cultural intermediaries Katz, Lavie, and Wolinitz), the market for products labeled as "green," "healthy," "fair," or "organic" is based on those who "believe in the subject," as Landes put it, those who realize their social and environmental care through everyday consumption practices. Thus, as Alkon and Guthman explain, this approach devolves regulatory responsibility from publicly accountable experts to consumers who can decide for themselves whether certain practices should be condoned or condemned.[90] Most of the participants involved in the process of formulating the Israeli law expressed the same approach. Thus, the law consisted of certification processes and ways of labeling produce as organic without discussing other substantial environmental and health issues or trying to use the legislative process to challenge, oppose, or change the Israeli conventional food system.

The discussions on the design of the law lasted about five years. Most of the conflicts between the lawmakers revolved around technical matters. For example, they debated at length the ways in which the organic label should be designed and they argued over which governmental organization should be responsible for the supervisory bodies, while taking for granted that the certification and supervision would be privatized—"as is happening in the United States and elsewhere in the Western world," as Yoram Porat, a representative of the IBOAA, said at the Economic Affairs Knesset Committee meeting mentioned above. None of the participants in the formulation of the law expressed a critical view of how the term "organic" was understood and translated conceptually: a food product that is supposed to be free from pesticides or chemical fertilizers and whose production costs are imposed on farmers and are therefore passed on to consumers by charging them premium prices.

Reforms in conventional agricultural and food systems were never brought up for discussion. Furthermore, the discussions throughout the legislative process did not raise the possibility of integrating into the Israeli organic law other non-conventional agricultural aspects, such as issues related to labor conditions or food justice—for example, the recognition and development of native Palestinian-Arab small holders' sustainable agricultural methods. A clear example of the localization and politicization of the organic can be seen in the ramifications of the Israeli Law on the Israeli-Palestinian conflict. Furthermore, in 2018, under political right-wing pressure, the Israeli Army approved the implementation of the Israeli Law for the Regulation of Organic Produce in the West Bank settlements.[91] It should be noted that although Israeli law does not cover those areas per se, the head of the Israel Defense Forces Central Command has the authority to decide to apply certain statutes there. Members of the opposition call such actions the "creeping annexation" of the Jewish settlement in the West Bank. Thus, in stark contrast to Wolinitz's original dream of the world's first organic state, the organic law serves as a means of strengthening Israel's rule of law in those contested settlements and thereby making a "green" or "organic" peaceful resolution to the conflict much more elusive.

To conclude, this chapter demonstrates how both cultural and institutional intermediaries translate global ideas related to the term "organic" according to a commodity-centric understanding of this notion and by assimilating it into a local cultural and political context. On the cultural side, Israeli cultural intermediaries tie the notion "organic" to a neoliberal understanding of sustainable consumption, to global "foodie" culture, and to global/Western health and natural foods movements and self-care.[92] In doing so, they promote an Israeli version of self-oriented global eco-habitus. This habitus stems from, and fortifies, cultural currents of post-Zionism and cosmopolitanism in Israel. On the organizational side, as reflected in the enactment of the Israeli Law for the Regulation of Organic Produce, institutional intermediaries contributed to the structural integration of Israeli agriculture in the global sphere, encouraged an export-oriented Israeli organic sector, and thus imposed neoliberal and post-Zionist ideologies on Jewish and Palestinian farmers. In addition, by establishing a legally recognized mechanism of control over the notion "organic" and over the process of obtaining organic certifications, the state demarcated Israeli organic agriculture within class and ethno-national boundaries and so contributed to discriminatory practices in access to sustainable and healthy agricultural produce.

Local culture, politics, and economy take on an important role in the vernacularization and implementation of socio-environmental ethical ideas. Like much of sociology, social anthropology, and critical cultural studies, the next chapter reminds us that positively valued symbols and ideas, such as organic food, are rarely as straightforward as they seem. Instead, they may be valorized by the prevailing social forces, which serve to amplify some aspects of these ideas over others. In Israel, "organic" has supported export-oriented agriculture, nationalism, global neoliberalization, and class and ethno-national distinction. This is not to suggest that Israelis (and others) striving for a more environmentally and socially just food system should necessarily eschew organic options. Rather, in response to Arjun Appadurai's call to describe a realm of possibilities,[93] the concluding chapter aims to advance the recognition of the local and global complexities of the meanings of organic, and to encourage Israeli farmers, producers, consumers, and distributers to continue seeking ways to enact their values not only through the production and consumption of labeled commodities but also through other collective means.

# Conclusion

## Glocal Organic

In October 1993, McDonald's opened its first branch in Israel. Soon, the chain became the focus of vociferous cultural and political public debates. Beyond the controversy, the debates made it clear that McDonald's presence was not merely a sign of fast food culture. Instead, to the Israelis of the mid-1990s, the brand was coupled with ideas related to Americanization and globalization.[1] Similar to in other places in the world,[2] McDonald's assimilation into Israel was a clear symbol of global culture. Seventeen years later, a program broadcast on Channel 2—at that time a popular Israeli commercial television channel—attested that McDonald's is no longer the only culinary signifier of globalization in the country. Journalists Guy Maroz and Orly Vilnai hosted this program, which included a televised investigative report. The program followed Maroz, who ate only organic food for a whole month to explore the impact of an "organic diet" on his health. "The idea was taken from Morgan Spurlock, the American man from *Super Size Me*, who ate only at McDonald's. We thought to ourselves—why not do the opposite and explore the impact of healthy food?" explained Maroz. The successful 2004 documentary movie that Maroz referred to enhanced and reflected the growing critique in the United States of fast food chains like McDonald's. Maroz and the producers of the Israeli documentary television show imitated the editing, soundtrack, and cynical-humoristic style of *Super Size Me*, albeit in this case their critique was aimed not at fast food, but at organic food—the supposed antithesis to the former, the staple of "the good food movement,"[3] which is considered contradictory to the global consumer culture.

At the end of a month in which Maroz's intense consumption of organic food was documented, the "experiment" was broadcast on *Orly and Guy Inc.* In the episode, Maroz explained the stages of the "experiment": "I, your loyal servant, did not touch anything that wasn't organically certified. The people who managed the experiment were very strict and allowed me to eat only organic products that were certified organic. What I ate can be described as 'organic strictly-kosher.'" Later in the episode, Maroz concluded that "organic food is not as healthy as people tend to think" and complained that he gained 2.5 kilograms (5.5 pounds) of body weight. He also complained about the high price of organic food. "Every purchase I made during this crazy month was recorded and assessed, and the findings, friends, are frightening. Prepare your wallets or talk to your bank-account manager, because you'll need to get a loan if you insist on eating organic." The show ended with the claim that the growing sector of organic food in Israel deceives its consumers. Maroz alluded that "whoever buys organic food is actually a *freier.*"[4] "The only difference between organic food and non-organic is the price. Whoever is willing to pay the price, that's his business," he declared, heaping scorn on the "organic culinary trend that swept the Western world and Israel," as he put it.

The show illustrates a mismatch between what is, allegedly, expected to be found in organic food and in the "organically certified" realm Maroz followed: health and personal benefits. The final twist of the show "revealed" that organic food can be fattening, "just like fast food," and that it can be expensive. Environmental or social issues were not the focus of the show. Instead, it illustrated the variety of culinary artifacts and trends that currently signify "cultural globalization" for the middle and upper-middle classes in Israel, the audience to whom the program was directed. It reflected on the ways in which alter-global ideas and images[5]—such as the anti-consumerist approach embedded in the documentary movie *Super Size Me*, or the philosophical ideas related to environmental and social change associated with the notion "organic"—can be imbued with global meanings. The analysis of organic food in Israel described in this book is meant to further our understanding of the interweaving of "global culture" with "alter-global culture." It points out that structural and cultural processes of globalization (usually associated with industrialized foods, with the Americanization of culinary culture, and with foods representing "ethnic" or "exotic" images) can be embedded in, and conveyed by, foods that are in and of themselves supposed to undermine the globalization and industrialization of food.

Previous studies have already shown both the advantages and the limitations of food movements in engaging with transformational shifts in the food system and in broader social and environmental realms.[6] The case of organic food in Israel—a focal point of the encounters between globalization and localization, between industrial farming and agro-ecology, and between hedonistic consumerism and ethical collectivism—extends this research by offering a *glocal* model for understanding "food for change." This model exposes the tensions that bind potentially radical concepts (such as "alternative food," "food activism," "food justice" or "eating for change") and reveals the limitation of how revolutionary they can be as they are caught up in cultural and political-economic forces of globalization, nationalism, and neoliberalization. It shows how the meanings of organic (and potentially similar foods) vary from place to place, even while they are also shaped by overarching global institutions and cultural frameworks.

By examining the manifestation of "alternative food" beyond the Global North, and by applying a sociological-*glocal* lens, we can better understand the roles of local culture, politics, and political economy in the vernacularization and implementation of global alternative ideas and practices (food-related or otherwise). As in other places, organic agriculture and organic food culture emerged in Israel accompanied by expressive-symbolic rhetoric and discourse proposing a fundamental alternative to the industrialization of food production as well as to the toxicity and environmental risks associated with the growth of agribusinesses. In Israel, organic food also emerged with the promise that "the taste of an organic cucumber is the taste of the Israeli cucumber of yesteryear, the taste that has been lost over the years and that we remember from the time when we were kids," as one organic farmer from Southern Israel told me. "Look at McDonald's," he said. "Now think the opposite—that is what organic food is all about. It is not industrialized, nor is it contaminated, like the produce coming from the [occupied] territories." However, examining organic food in Israel from a structural-institutional perspective exposes that no straightforward distinction between organic and conventional food has ever existed. As alluded to many times throughout the chapters of this book, the boundaries between "global, industrial, high-tech and conventional agriculture" and "alternative farming" are drawn in complex, locally contextual ways.

Drawing upon previous research that opposed the conventional/organic binary,[7] this book points to moments when collaboration and alliances between "conventional" and "organic" not only existed but also dialectically evolved. The Israeli case exposes the dynamics between the conventional and organic food sectors as they move from a mutually beneficial alliance to "conventionalization" and, at times, toward "bifurcation of the organic"—that is, the divide of the field of organic agriculture into large-scale growers using industrial practices and small-scale food producers and farmers tending towards more agro-ecological methods.[8] Unlike previous research, this book moves beyond the structural, institutional, and political-economic levels to apply a cultural analysis that examines the tensions and paradoxes of "alternative foods" through the relations between the global and the local as a composite of the structural and symbolic levels.[9] Drawing from the expressive-symbolic level, this case illustrates how countercultural "counter-cuisine" can also be seen as mainstream "global cuisine"[10] and how "cosmopolitan tastes" and the so-called "foodie's ethical/political taste"[11] are found not only among the more liberal, secular, and urban consumers traditionally associated with alternative food movements, but also among groups with secluded, ethnocentric, or parochial worldviews. Ultimately, the case of organic food in Israel exposes the "gray areas" where a given field and its alternatives are intertwined—both on the structural and symbolic level.

## The Vernacularization of Organic Food in Israel

Except for the recent trend of veganism that has swept parts of Israeli society since 2010 (an important topic of sociological research that requires a separate analysis),[12] organic food has been the only agro-food category in which discursive (if not substantive) forms of heterodoxy (or resistance) were deliberately proposed in opposition to the orthodox field of food and agriculture in Israel (to use Bourdieu's field theory terminology[13]). Fair Trade, for example, was never discussed and exercised beyond a small group of advocates, a few (mostly Palestinian) food producers, and food activists.[14] Despite a few attempts, Fair Trade has not been assimilated institutionally or discursively in Israel. Local food and the culture of locavorism,[15] to take another example, have been conceived (mistakenly!) by many consumers as almost laughable in a small country such as Israel. This is despite the fact that since the 1990s the Israeli government has implemented policies encouraging the import of food products to Israel and Palestine.[16] Regardless of a few efforts to draw public attention to the social and environmental importance of eating local,[17] these issues have not been on the cultural agenda in Israel and never developed into a widely exercised practice.

Organic food, by contrast, has developed into a vibrant field in Israel. Many in Israel—especially organic food producers, consumers, and distributors, who are all endowed with high economic and cultural capital—associate the notion organic with aspects of health, sustainability, locality, and care. However, many of them tend to adopt such values in a narrow way and mostly exercise them on the discursive-expressive level. The realization of the "organic ideal," namely the constitution of a field that is based on agro-ecological practices, sustainable and fair modes of production and distribution, and ethically driven consumers, does not hold up empirically to the realities of the field of organic food in Israel and Palestine.

In an attempt to provide some answers to the questions that have guided this book—how and in what ways has organic food been realized and assimilated in this region—the analysis presented here indicates that "organic" has never had a unified set of agrarian and marketing roles that were simply assimilated in a local context but is rather derived from a competing set of meanings[18] and from varying practices constructed by global cultural and structural conditions. At the same time, a certain set of fundamental local socio-political and cultural aspects of Israeli society underlined the emergence and development of "organic," allowing and limiting its cultivation in Israel and Palestine.

As I have shown in my history of the field of organic food in Israel, since its inception it has stood side by side with conventional agriculture. Both conventional and organic agriculture were drawn from the same cultural and ideological sources. This is evident from the endeavors of Mario Levi and his colleagues from the IBOAA, who identified organic agriculture as a new vehicle for the Zionist project that had long been propelled by conventional agriculture. As these "organic pioneers" were themselves members of the Zionist elite group, they tied the emerging field of organic food to the core of Israeli identity and thus won recognition and gained legitimization and support from the (then) state-supported conventional agriculture. It should not be surprising, then, that the founders of organic farming in Israel have always been reluctant to offer critique of the national-cum-global-neoliberal trajectory of Israeli conventional agriculture. On the contrary, they have been pleased with the institutional recognition they have received and they have contributed to the strengthening of Israeli (Jewish) conventional agriculture as well as to its transformation into a global-oriented agribusiness. Thus, in the formative years of the establishment of organic food in Israel, actors in the field of organic food operated with global orientation and contributed to the overall global trade of Israeli agricultural produce.

Blunt resistance to this export-oriented direction, if there was any, was eradicated—or at least did not reach the core discourse in the field of organic food. Instead of sticking to the philosophical ideas of organic ("locality, being true to ecological regions, social justice, and equality"[19]), many of the organic farmers were drawn to neoliberal ideas, suggesting that "export is crucial to the survival of organic [agriculture]. Without export, organic agriculture in Israel would not exist at all, definitely not in terms of large quantity, high quality and variety of produce," as goes the "mantra" that is frequently used during the IBOAA's course "Basic Organic Agriculture," a mandatory course for organic farmers in Israel.[20]

But the focus of Israeli organic agriculture in its formative years was not only on crossing borders and seas. During that time, it also engaged in local meanings of cultivating the land and with local territorial boundaries. In this regard, one of the most striking manifestations of the vernacularization of organic agriculture in the local Israeli/Palestinian context is related to one of the fundamental elements of the organic philosophy: the soil. Many of the organic farmers I interviewed recited the basic premise that "the organic philosophy began as a philosophy of the soil, and soil quality was the first test of organic affiliation."[21] It was exactly this philosophical principal that enchanted neo-Zionist Jewish settlers who adopted it as a means of developing the settlement movement in the Gaza Strip and the West Bank and for reclaiming their indigeneity.

It was only after 2005 that a domestic trade of organic food emerged and a local Israeli field of organic food began to operate. From the structural level of production, several actors in the domestic field (farmers, food producers, distributors, cooks, marketers, and organic advocates) started to direct their endeavors toward gaining both status and position, not only in the field of organic food or in the nutritional-medical fields but also by moving from the margins of the broader Israeli agricultural and culinary fields into the mainstream. As such, many actors in the field of organic food never developed a critical standpoint (whether they ever did try, or will try, remains an open question). Gradually, they were assimilated into the larger "fields of power"[22] according to which the agricultural and the gastronomic fields operate: the Israeli (globalized and neoliberalized) political-economic field and the cultural-political field. We have seen, for example, how both structural globalization and neoliberal political-economic ideology in Israel directed the import of growth-oriented organic farming methods, standards of production, global regulatory methods, codes of certification, and global/American retailing methods.

On the cultural front, globalization is clearly influential. All actors in this field, most of whom demonstrate eloquence in the global cultural discourse associated with transnational elites,[23] display their cosmopolitan identities through organic food consumption, production, or distribution. Through the mediation of organic food, many of them demonstrate and reflect on their previous or anticipated engagements with the globe, cultivating varied forms of global habitus in the process.[24] Thus, the field of organic food in Israel is shown to be made up of a multitude of discursive and performative frames of "the global": "American" lifestyle and global consumer culture; New Age culture; urban culture; Western dietary advice;[25] the global discourse of environmentalism;[26] global therapeutic and self-care narratives that fit an emotional discourse typical to the precariat in the age of neoliberal capitalism[27] ("I switched from high-tech and found my true Self," was recounted by many organic farmers I spoke to); global foodie culture;[28] neoliberal discourses of ethical consumerism that obfuscate responsibility to the environment and distant others;[29] and more. Other actors in this field, such as the Jewish settler-farmers in the occupied territories, use the global meanings associated with the notion organic to promote—paradoxically—a local-parochial lifestyle, to gain legitimacy for discriminatory (or illegal) acts, or as a foothold for accessing a broader domestic retail market and even global trade.

As such, organic food remains, in the Israeli/Palestinian context, an issue of global/local orientation—a culinary category that some people use as a local food that supports their global identity and that others use as a global food that supports their local identity.

## Adoption-Translation-Assimilation and the Rights for Better Food

Occurrences of aesthetic-cosmopolitanism—that is, as defined by sociologist Motti Regev, cultural openness to global styles and contents in a way that is believed to express local (national, ethno-national, and communal) uniqueness[30]—are noticeable in the Israeli field of organic food. One emblematic example is ascribing a global cultural meaning of "organicness" to "the national dish" (hummus), to go back to the opening vignette in the introduction to this book. Forms of global-local interconnectedness, as well as connectivity between local cultures and global technologies and media, which also typify aesthetic-cosmopolitanism, are also evident.[31] These can be discerned in the adoption of the branding and marketing methods used to sell local and organic foods (see Eden Teva Market, Chapter 4); in the use of global communications networks and social media as primary sources of knowledge and networking between CSA consumers and producers (Chapter 3); in the ways in which consumers repeatedly describe the act of shopping for organic food as a "phenomenological experience of being abroad"; in consumer-producers' attachments to global foodie culture; and in global trends of using artisanal quality foods as a means of articulating a "new Israeli cuisine" (for example Ronit Vered, see Chapter 5). Global-local connectivity is even exercised by those who use organic food to promote a neo-national political culture (i.e., organic settler-farmers, see Chapter 2).

Similarly, the agricultural means of production and the organic foods themselves attest to the dominance of aesthetic-cosmopolitanism in this field. We see this in the ways in which organic ingredients imported from the four corners of the world are combined with foods embedded with national significance (imported organic sesame, chickpeas, and whole wheat used to make an organic hummus meal, for example). We see this in how herbal and mineral additives prepared by biodynamic research and development centers in Europe are used by Israeli kibbutzim. We also see this in the way in which herbs that are perceived as exotic share the same beds with vegetables and legumes that have been grown in Middle Eastern soil for centuries. And we see this in the ways in which Israeli organic retailers have adopted "American" management styles, marketing techniques, and décor in trying to create the same "shopping experience" found in Global North outlets (such as Whole Foods Market or Trader Joe's). One can find on these aspirant-American organic supermarkets' shelves bottles of organic olive oils imported from Italy alongside bottles of organic olive oil manufactured in Neot Smadar (a kibbutz in Southern Israel where many of its members make their living from local organic farming).

Ultimately, patterns of "adoption-translation-assimilation" characterize the glocal practices and culinary artifacts described above. In effect, these patterns are revealed as the doxa of this field.[32] That is to say that the taken-for-granted engagements, with "wider shores of cultural experience,"[33] are shared by all actors in the Israeli field of organic food. As such, practices of global cultural-translation serve as the cornerstone of this field, and the "organic" in Israel has been imbued with the symbolic power of cosmopolitanism, no less than the symbolism it holds as healthy, ethical, or environmental food.

Alongside the seeming uniqueness of the Israeli field of organic food that binds together the notion of "organic" with the "global," this field also shares some important similarities to fields of organic food in the Global North. Research done in the US has pointed out that it became commonplace for the professional middle class—disproportionately white, educated, and relatively young—to reject mass-market industrial products and turn to "alternatives" (organic food or local food, farm-to-table, among other gastronomic alternatives).[34] We have witnessed similar tendencies in the field of organic food in Israel, which is demarcated by the boundaries of privileged groups in Israeli society: upper-middle-class Ashkenazi Jews, new middle-class Mizrahi Jews, Jewish-Americans, the newly religious former moshavniks (members of a moshav) who have settled in the occupied territories, veteran halutzim ("pioneers"—Zionist ideological farmers and workers), and Zionist kibbutzniks (members of a kibbutz). These are the "citizen-consumers"[35] who have access to organic food and can afford it.

The notion "citizen-consumer" is often related to organic food consumers. It is a concept that combines environmental responsibility and the reduction of social inequality on the one hand with quality, taste, naturalness, and health on the other hand.[36] However, if we take into account the assumption that food-related issues are not always and not necessarily exempt from "real" politics,[37] organic food can be used as a lens for understanding the complexities of citizenship-consumerism in Israel and Palestine as well as issues related to "eating rights" in this region (namely, the accessibility of healthy and sustainable food). It exposes who can participate as a "food citizen"[38] and, more broadly, the limitations of practicing sociocultural citizenship in Israeli society via mundane practices such as food production and consumption. Tracing the sociological structural and symbolic levels of organic food in Israel reveals those who have been granted protection from the ills of the "dysfunctional modern (and conventional) food system."[39] It also delineates who has the ability to "vote with their money" for better food. Not surprisingly, and in line with the "unbearable whiteness of alternative food in the US," to use Julie Guthman's argument,[40] those who are at the margins of Israel's globalized political culture are the ones who are left out. Members of the lower classes, foreign workers, Palestinian farmers, unemployed single mothers—these Israeli residents may find themselves as hard-labor workers in the fields owned by an export-oriented organic Jewish farmer in the Arava desert or working as manual labor in an organic tahini production factory located in the West Bank or even working in some of the CSA initiatives, but they are not included in the privileged groups of citizen-consumers, and thus their access to safer and healthier food is limited.

As the philosophical foundations of organic agriculture moved across geographical and cultural boundaries, they became entangled with prominent local issues. In the Israeli case, organic food has been imbued with some of the most salient historical, cultural, political, and economic issues: the ethos of halutzim, the utopian visions of the Israeli kibbutz, the indigeneity and land ownership claimed by Jewish settlers in the Gaza Strip and the West Bank, and also the Americanization of Israeli society and its neoliberal economy. Nevertheless, despite these local meanings and regardless of the widespread public discourse that organic agriculture has provoked, it is inextricably connected to the globalization and commodification of food in Israel. Therefore, the expansion of the field of organic food in Israel has replicated what it was set up to oppose, and it has failed in proposing alternative ways to cope with climate change or address problems of food safety, food security, and social justice.

The case of organic food in Israel reflects some of the most pressing questions concerning Israel (as well as other countries in the "post-global" era, after the 2016 USA presidential elections, the UK Brexit vote, and the COVID-19 pandemic crisis) regarding globalization, nationalism, and neoliberalism. It uncovers the ways in which contradicting, and at times complementary, tendencies—such as environmentalism and colonialism, cosmopolitanism and (neo)nationalism, spiritualism and market-driven materialism—influence and are affected by macro-level structural processes as well as spheres of everyday life, such as farming, cooking, or eating.

# Notes

## Introduction

1. Vered, 2015.
2. See, for example, Gvion, 2012; Hirsch, 2011; Ranta, 2015.
3. Ortner, 1973.
4. Ariel, 2012; Avieli, 2016; Hirsch & Tene, 2013.
5. Hirsch & Tene, 2013, Ichijo & Ranta, 2016. For more about gastro-politics and gastro-nationalism, see DeSoucey, 2016; Bégin, 2016.
6. Grosglik & Ram, 2013.
7. Appadurai, 1986.
8. "Israel" refers to the State of Israel within armistice lines demarcated in 1948 (also known as the Green Lines). "Palestine" refers to the West Bank and the Gaza Strip.
9. Raviv, 2003.
10. The Green Line refers to the armistice line following the 1948 War, and the territories beyond the Green Line are those conquered by Israel in the 1967 Six-Day War and currently under military and administrative rule.
11. For example, the definition of the notion organic that has been offered by the International Federation of Organic Agriculture Movements (IFOAM) says that "organic agriculture should build on relationships that ensure fairness with regard to the common environment and life opportunities." See IFOAM, n.d.(b).
12. Lavie, 2009.
13. Kaufman, 2007.
14. Heelas, 1996; Liechty, 2017.
15. Johnston, 2008.
16. Ritzer, 2004.
17. Robertson, 1995; J. Watson, 2006.
18. Holmstedt, 2010, p. 111.
19. This explanation is taken from a widely read article by renowned Israeli author Meir Shalev, called "The Hummus Is Ours" (2001).

20. For more about the notion "gastropolitics," see DeSoucey, 2016, pp. 14–19.

21. For more on organic hummus, see Grosglik, 2011.

22. Julier, 2013, p. 4.

23. The English agronomist Sir Albert Howard, the Austrian philosopher Rudolf Steiner, British farmer Eve Balfour, agriculturalist Walter Northbourn, American farmer and environmental activist Wendell Berry, publisher Jerome Irving Rodale, and many others are among the champions of organic farming. See Duram, 2010, pp. xv–xxxi; Voget, 2007.

24. Barton, 2018, p. 17.

25. Barton, 2018, p. 12.

26. Belasco, 2007; L. J. Miller, 2017, p. 18.

27. Haedicke, 2016, p. 7.

28. Bivar, 2018, pp. 1–12.

29. Lévi-Strauss, 2012 [1968]; see also Shapin, 2006.

30. Eden, 2011; Raynolds, 2004.

31. Johnston, 2008; Baumann et al., 2017.

32. See IFOAM, n.d.

33. IFOAM, 2018, p. 68.

34. Guthman, 2004.

35. See, for example, Buck et al., 1997; Guthman, 2004; Jaffee & Howard, 2010.

36. Goodman et al., 2012, p. 153; Haedicke, 2016; Obach, 2015.

37. Guthman, 2004; Haedicke, 2016, p. 6.

38. Johnston et al. 2009; Pollan, 2006; Guthman, 2004b; Fromartz, 2006.

39. Johnston & Baumann, 2009; Johnston, 2008.

40. Johnston, 2008.

41. Johnston & Szabo, 2011; Johnston et al. 2011; Cairns et al. 2013.

42. Guthman, 2003; Alkon & Guthman, 2017; Alkon, 2012; Guthman, 2011; Johnston & Szabo, 2011; Goodman et al., 2012.

43. Johnston, 2007; Guthman, 2003.

44. Goodman et al., 2012; Guthman, 2007; Johnston & Cairns, 2012.

45. Alkon & Guthman, 2017, p. 12.

46. Guthman, 2007. For more about the counter-critique of this critique, see Alkon & Guthman, 2017, pp. 15–18.

47. One exception is Guntra Aistara's *Organic Sovereignties* (2018), which provides an ethnographic description on the identity-making of organic farmers in Latvia and Costa Rica.

48. Inglis & Gimlin, 2009; J. Watson & Caldwell, 2005; Phillips, 2006; Wilk, 2006.

49. Gastro-anomie is a term coined by sociologist Claude Fischler (1979); it refers to the industrialized, individualized, chaotic, and accelerated foodways that characterize modern societies.

50. Raynolds, 2000.

51. Ritzer, 2004; Schlosser, 2001.

52. For more about science and "risk society" as well as on risk, science, and food, see Latour, 1988; Paxson, 2012.

53. Orlando, 2018.

54. Beck, 1992, p. 39.

55. Giddens, 1990, p. 21.

56. Pleyers, 2010; Leitch, 2012.

57. Barton, 2018, p. 156.

58. Barton, 2018, p. 192; Geier, 2007.

59. Barton, 2018, p. 183.

60. Barton, 2018, p. 191; Obach, 2015.

61. Barton, 2018, pp. 184–203.

62. IFOAM, 2018, p. 13.

63. IFOAM, 2012, p. 27.

64. IFOAM, 2018, p. 21.

65. Barton, 2018, p. 195.

66. IFOAM, 2018.

67. Tovey, 2009.

68. Warde, 2016.

69. Appadurai, 1986.

70. R. Gross, 2019; see also MARD (n.d.).

71. R. Gross, 2019; see also MARD (n.d.).

72. Yurista, 2015; Herskovitz, 2011.

73. Kahal, 2007.

74. Yurista, 2015.

75. Herskovitz, 2011.

76. Mazori, 2005.

77. Bar-Zuri & Warshevski, 2010.

78. Raviv, 2015; Avieli, 2017; Rozin, 2006.

79. Gutkowski, 2018.

80. Ranta & Mendel, 2014.

81. Ram, 2008.

82. Grosglik & Ram, 2013.

83. Katz-Gerro, 2009.

84. DeSoucey, 2016.

85. Weiss, 2016.

86. The term *habitus*, as suggested by the French Sociologist Pierre Bourdieu, refers to the dispositions, to the schemes of perception, and to appreciation of practices that individuals embody and possess as well as to "the mental structures through which they apprehend the social world." (Bourdieu, 1989, pp. 18–19; see also 1972/1977; 1984.)

87. See, for example, L. J. Miller, 2017; Ray, 2016; DeSoucey; 2016; Johnston & Baumann, 2009; Ferguson, 1998.

88. Martin, 2003; Bourdieu, 1993, p. 72.

89. Sewell, 1992; Regev, 2007.

90. Fligstein & McAdams, 2012, p. 10.

91. Johnston & Szabo, 2011.

92. Senor & Singer, 2009.

93. Ferguson, 2011, p. 378.

94. See, for example, Guthman, 2004; 2014.

95. See, for example, Haedicke, 2016; Obach, 2015.

96. See, for example, Johnston & Szabo, 2011; Cairns et al., 2013.

97. Simmel, 1915/1997.

98. A field, as defined by Bourdieu, is "[a] structured space of position (or posts [and relations between actors]), whose properties depend on their position within [this space]" Bourdieu, 1984/1993b, p. 72; Bourdieu 1990, p. 192.

99. See also L. J. Miller, 2017, p. 19.

100. For more about the distinction between the structural-institutional and the symbolic-expressive levels of the societal, see Ram, 2004; Rosa, 2004; Schudson,1989; DiMaggio & Powell, 1983.

# Chapter 1

1. Eisenstadt, 1967; Sternhell, 1998; Kellerman, 1993.

2. Tal, 2007; Kamen, 1991; Abufarha, 2008; Zerubavel, 1996, Shani, 2018; McKee, 2016.

3. S. Almog, 2000, p. 91; Hirsch, 2015; Shapira, 2012, pp. 42–54.

4. Halpern & Reinharz, 1998, pp. 186–197; Shafir & Peled, 2002, p. 39.

5. Krampf, 2018; Swedenburg, 1990; Kamen, 1991; Zaban, 2012.

6. Kimmerling & Migdal, 1993; Tal, 2007.

7. In accordance with Engel's law, which states that as income rises, the proportion of income spent on food is reduced even though the absolute expenditure on food rises. See Zimmerman, 1928.

8. Kislev, 2015; Schwartz, 1995.

9. Shalev, 2000; Ram, 2008.

10. Kislev, 2015; Schwartz, 1995; Tal, 2007.

11. Levi, 2016, p. 15.

12. The name of the Ministry of Agriculture was changed in 1992 to Ministry of Agriculture and Rural Development, hereafter MARD.

13. Lockeretz, 2007, pp. 18, 152–154.

14. Levi, 2016, p. 9.

15. Neumann, 2011, p. 5.

16. Levi used key phrases from the culture and ethos of Jewish-Zionist society, referring to the contradictory challenge Israeli society faced: striving to simultaneously be a nation like all others as well as a "light unto the nations" (see Brenner, 2018).

17. Bio Tour, n.d.

18. Kibbutz Sde Eliyahu, n.d.

19. Carson, 1962; Montrie, 2018.

20. See, for example, Guthman, 2004; Obach, 2015; Haedicke, 2016; Bivar, 2018.

21. Obach, 2015, p. 57; Lockeretz, 2007, p. 175–186.

22. Finkelshtain & Kachel, 2006.

23. Israel Port Company, 2009.

24. Buck et al., 1997; Guthman, 2004.

25. Zali, 1993.

26. Kornet, 1993.

27. Israel Port Company, 2009.

28. Levi, 1993.

29. De-Vinter, n.d.

30. Bourdieu, 1969; Ferguson, 1998.

31. Swidler, 1986.

32. Adler, 2005: 23.

33. "Gastronationalism" refers to the ways in which materialistic and symbolic practices related to foods and food-related issues are used to promote a nationalistic sentiment with political overtones (DeSoucey, 2016, p. 32; see also Bégin, 2016).

34. The Balfour Declaration was a statement issued by the British government in 1917 announcing support for the establishment of a "national home for the Jewish people" in Palestine. The declaration was sent from the then foreign secretary of the United Kingdom, Arthur Balfour, to Lord Rothschild (see Shapira, 2012, pp. 70–77). Eve Balfour was the niece of Arthur Balfour.

35. Fine & DeSoucey, 2005; Fine & Wood, 2010.

36. This refers to Rabbi Yoel Moshe Salomon, one of the founders of the Jewish settlement in Palestine, beginning what is known as "the mother of rural Jewish settlement," Petah Tikva (1878).

37. Paltin, 1996.

38. Shapira, 2012, pp. 58–59; see also Shapira, 2004.

39. Adler, 2005, p. 21.

40. Shenhav, 2007; Ram, 2011.

# Chapter 2

1. Arad, 2012.

2. "An organic fraud in Gush Katif," 1997.

3. Methyl bromide is a chemical substance used as a fumigant to disinfect soil and protect against pests during food storage and shipment. Israel is one of the main manufacturers of methyl bromide in the world. Since it is known as

an ozone-depleting substance, the Montreal Protocol banned the use of methyl bromide in 183 countries, including Israel. For more about the social politics of methyl bromide and post–methyl bromide fumigation, see Guthman, 2016.

4. Douglas, 1966/1978.

5. For more about the Israeli disengagement from Gaza and the withdrawal of the Israeli army from inside the Gaza Strip, see Dromi, 2014.

6. Schnell & Mishal, 2005, p. 34.

7. Feige, 2009; Ophir et al., 2009; Shafir, 2017; Allegra et al., 2017.

8. As defined by the IFOAM General Assembly in Vignola, Italy, in June 2008. See IFOAM, 2008.

9. Gieryn, 1983; Habermas, 1970; Gouldner, 1976.

10. See more about attributing spiritual meanings to organic food in Chapter 4.

11. Conford, 2001; Reed, 2001.

12. Barton, 2018.

13. Barton, 2018, pp. 2–3.

14. Sayre, 2011.

15. Reynolds, 2015. For more about the agro-economic realities of Palestinians in the West Bank, see Abu-Sada 2009; Algazi, 2009.

16. Etkes, 2013.

17. Gush Emunim was an Israeli Orthodox Jewish right-wing activist movement committed to establishing Jewish settlements in the West Bank, the Gaza Strip, and the Golan Heights. For more on Gush Emunim, see Feige, 2009, pp. 26–29.

18. Taken from the MARD report (2005).

19. M. Drori, 2005.

20. In the Israeli context, the term "Anglo-Saxons" is used to describe the Jews who immigrated to Israel from United States, UK, Canada, Australia, New Zealand, Ireland, and South Africa. For more about the wave of settlement in the territories captured by the State of Israel during the 1967 War by Jewish "Anglo-Saxons" (among them Jewish Americans), see Hirschhorn, 2017.

21. Jasanoff, 2005, p. 7.

22. Jasanoff, 2004; 2012. For more about science and agriculture in Israel/Palestine, see Gutkowski, 2018.

23. Ram, 2008.

24. Arad, 2012.

25. Shragai, 2007.

26. Adler, 2005, p. 143.

27. Feige, 2009, pp. 234–235; Haklai, 2007; Sasson, 2005; A. Gross, 2017.

28. Har-Sinai, 2020.

29. Beyond the Lines, 2018.

30. Har-Sinai, 2020.

31. Beyond the Lines, 2018.

32. Feige, 2009, pp. 28–29.

33. Taken from Beyond the Lines, 2018.

34. Even-Zohar, 1981; Zerubavel, 2008; Hirsch, 2016.

35. Tesdell, 2015.

36. Karish-Hazoni, 2016.

37. Karish-Hazoni, 2016.

38. Har-Sinai, 2020.

39. Taken from Beyond the Lines, 2018.

40. Taken from Meshek Achiya, n.d.

41. For more about the contemporary Israeli olive oil sector, see Simovitch, 2015; Gutkowski, 2016.

42. Taken from Ventura, n.d.

43. Monterescu, 2017. See also Trubek, 2008; Trubek et al., 2010.

44. Bonacich, 1972.

45. See also Shafir, 1996.

46. Karish-Hazoni, 2016.

47. Feige, 2009, pp. 234–246.

48. Feige, 2009, p. 237.

49. Fischer, 2011.

50. Fischer, 2011, p. 302.

51. Karish-Hazoni, 2016.

52. Strained yogurt in Arabic. Labneh is widely eaten in the Levant, Egypt, and the Arabian Peninsula. In Israel it is mostly considered an Arabic food.

53. Feige, 2009, p. 236.

54. See more about Lavie in Chapter 6.

55. Lavie, 2003b.

56. Lavie, 2003b.

57. *King James Bible*, 1611/2004.

58. Halevi, 2005.

59. Halevi, 2005.

60. Lavie, 2003b. See also Arad, 2012.

61. Halevi, 2005.

62. Lavie, 2003.

63. Lapid, 2010.

64. Brener, 2018.

65. Feige, 2009, p. 236.

66. Kotef, 2017.

67. Lavie, 2003b.

68. Feige, 2009, p. 237.

69. Feige, 2009, p. 238.

70. Arad, 2012.

71. Arad, 2012. See also Gutkowski (2018) for more about the construction of Palestinians as essentially primitive instead of seeing their agricultural practices

as an outcome or manifestation of the power relations that the Palestinian-Arabs are entangled with.

72. Gutkowski, 2010; Meneley, 2007; 2011; 2014.

73. Gutkowski, 2010.

74. Guthman, 2011b.

75. Karish-Hazoni, 2016.

76. Fischer, 2007; Feige, 2009, pp. 234–246.

77. Inbar & Osman, 2018.

78. In Israel the dairy market is operated according to a state production quota system, which is regulated by law.

79. Palti, 2007.

80. Palti, 2007.

81. Shragai, 2007.

82. Palti, 2007.

83. Handel et al., 2015; Monterescu, 2017.

84. Passidomo, 2017. See also Chapple-Sokol, 2013.

85. Handel et al., 2015.

86. Eldad & Bashan, 2011.

87. Eldad & Bashan, 2011, pp. 252–253.

88. Arad, 2012.

89. Shragai, 2007.

90. Arad, 2012.

91. Emerich, 2011.


# Chapter 3

1. Laor, 2017, p. 277.

2. Ilbery & Maye, 2005. See also Renting et al., 2003.

3. See, for example, Holloway & Kneafsey, 2000; Guthman, 2003; C. Morris & Buller, 2003; Hinrichs, 2000; 2003.

4. Renting et al., 2003.

5. As Illouz & John explain, the notion of "global habitus" refers to a cognitive and emotional disposition to move easily and smoothly from one national context to another, to quickly adapt and adopt different cultural outlooks, and to think of [oneself] as an agent operating in the world as a single unit. (2003, p. 211).

6. Galt, 2013; Cone & Myhre, 2000.

7. Guthman, 2004b; Galt, 2013; Goodman, 2000; Morgan et al., 2006; Haedicke, 2016.

8. Kloppenburg et al., 1996.

9. Morgan et al., 2006; Johnston, 2008.

10. Galt, 2013; Cone & Myhre, 2000.

11. Galt, 2013.

12. For example, *Time* magazine's article titled "Eating Better than Organic" explored "the new 'local foods' movements." See Cloud, 2007.

13. For more about mallow in the Israeli-Palestinian context, see Levit, 2018.

14. B. Morris, 2008.

15. Sorek, 2012.

16. Sorek, 2008, p. 60.

17. Bauman, 1998; Calhoun, 2002.

18. Hannerz, 1996; Brooks, 2000.

19. Bourdieu, 1984, p. 359; Lash, 1990; Ram, 2008, p. 76 (respectively).

20. Williams, 2010, pp. 10–16.

21. Hardt & Negri, 2004.

22. Florida, 2002; Clifton, 2008; Boltanski & Chiapellio, 2005.

23. Florida, 2004.

24. Lloyd, 2010; Leadbeater & Oakley, 1999; Scott, 2017.

25. Scott, 2017; Schiermer, 2014.

26. Scott, 2017.

27. Pratt & Hutton, 2013; Scott, 2017; Ocejo, 2017.

28. Illouz & John, 2003.

29. U. Cohen & Leon, 2008. It should be noted that class, ethnicity, and ethno-nationality are interrelated in Israel. In the Israeli upper middle class one can find a disproportionate number (compared to the lower classes) of Ashkenazi Jewish males. See Dagan-Buzaglo & Konor-Atias, 2013, p. 18.

30. Conscription exists in Israel for Israeli citizens over the age of 18 who are Jewish, Druze, or Circassian. Palestinians who are citizens of Israel are not conscripted.

31. E. Cohen, 2003; Uriely et al., 2002.

32. Bloch-Tzemach, 1998.

33. Chevannes, 1994.

34. Senor & Singer, 2009.

35. For more about the controversial history of the cherry tomato, see Wexler, 2016.

36. Boltanski & Chiapello 2005; Gill & Pratt, 2008; Urciuoli, 2008.

37. Hardt & Negri, 2004.

38. Illouz & John, 2003.

39. For a sociological account on the cultural phenomenon of the longing for the "taste from the past," see Jordan, 2015.

40. For more about the place of "the family" in contemporary Jewish identity and its relation to holidays, see Shoham, 2017, pp. 20–63.

41. For more about the concept of "the invention of tradition," see Hobsbawm, 1983.

42. Herzfeld, 2016, p. 139.

43. See also Caldwell, 2006, p. 106.

44. Habermas, 1981; Taylor, 1992; Banet-Weiser, 2012; MacCannell, 1976.
45. Goffman, 1959.
46. For more about artisanal work and the social construction of distinguishing artisan from commodity production, see Paxson, 2010; 2012.
47. Bourdieu, 1993, pp. 30–40, 51.
48. Bourdieu, 1993; Kennedy, 1998; Rothenberg & Fine, 2008.
49. Peterson & Kern, 1996; Peterson, 2005; Warde et al., 2007; Garía-Álvarez et al., 2007; Goldberg et al., 2016.
50. Granovetter, 1985; Uzzi, 1996; Hinrichs, 2000; Barham, 1997.
51. U. Cohen & Leon, 2008.
52. DeVault, 1991, p. 79; Lupton, 1996, p. 39; Szabo & Koch, 2017.
53. Ashley et al., 2004, p. 134; Yates & Warde, 2017.
54. Beck, 1992. See also Beamish, 2015.
55. Cf. Cairns et al., 2013; 2014.
56. For more about global subjects and global subjectivities, see Pfeifer, 2019; Bauman, 1998.
57. For more about the ways in which "heirloom tomato" emerged as a cultural object, transitioning from something grown by individual gardeners into a status symbol, in the late twentieth century, see Jordan, 2007; 2015.
58. For more about the citizen-consumer hybrid, see Johnston, 2008; Grosglik, 2017.
59. Alkon, 2008; Gvion, 2017; S. Watson, 2009.
60. Gvion, 2017. See also Duruz, 2004; Tiemann, 2008.
61. Ram, 2008, p. 71.
62. Gvion, 2017.
63. Stephenson, 2008; Robinson & Farmer, 2017, pp. 25–54.
64. Alkon, 2012, p. 22.
65. Giddens, 1990.
66. Guthman, 2008; Alkon, 2012.
67. For more about *MasterChef Israel*, see Grosglik & Lerner, 2020.
68. From Tel Aviv Port Market website, Shuk Hanamal (n.d.).
69. "Autumn in the farmers market" (no author mentioned), *Al Hashulchan* magazine, 2011.
70. Baudrillard, 1994.
71. Veblen, 1899/1934.
72. Johnston & Baumann, 2009.
73. For more about the proliferation of ethnic and specialty foods in Israel, see Grosglik & Ram, 2013; Kaplan, 2013; Hirsch, 2011.
74. Volkomir, 2010.
75. Coles & Crang, 2011.
76. T'zvi, 2010.
77. Swidler, 1986; Johnston & Baumann, 2009; Johnston et al., 2011.

78. Florida, 2002; Brooks, 2000; Schiermer, 2014; Veblen, 1899/1934.
79. Zukin, 2008.

## Chapter 4

1. Marx, 1857/1973.
2. Marx, 1857/1973, p. 94.
3. Marx, 1857/1973, p. 95.
4. Nitzan & Bichler, 2009; see also Howard, 2016.
5. Dixon, 1999; Trivette, 2019.
6. Goodman et al., 2012.
7. Obach, 2007; Obach, 2015; Howard, 2009.
8. Guthman, 2004; Obach, 2015; Haedicke, 2016; Johnston et al., 2009; Fromartz, 2006; Raynolds, 2004.
9. Johnston et al., 2009.
10. See, for example, Thompson & Coskuner-Balli, 2007. See also, about the consequences of the 2017 Amazon–Whole Foods deal, Dewey, 2017.
11. Johnston et al., 2009. See also Oosterveer & Sonnenfeld, 2011, pp. 203–205.
12. Zukin, 2008. See also Johnston & Szabo, 2011.
13. Polanyi, 1957; Granovetter, 1985; Zelizer, 2011; Zukin & DiMaggio, 1990, pp. 1–36.
14. Kellerman, 1993.
15. For more about Americanization in the Israeli context, see Ram, 2008.
16. Scholar Tamar Berger (2019) defines suburban places as "locus of the middle-class" and explains that in the Israeli context the suburbs can be identified as "neighborhoods built since the 1980s, both single-family detached houses and saturated construction, on city outskirts or at some distance from metropolises, places that sometimes have a certain affinity with cities although they are not totally dependent on them, places of the middle-class and fundamentally family-oriented places with specific lifestyles." See also Allegra, 2013.
17. The term New Age, or New Age Movement, refers to spiritual practices and a fragmented array of groups characterized by the anticipation of a spiritual cosmic transformation, the search for a spiritual life, the use of various mind-body techniques, and a discourse that extols the Self (through marketplace consumption) and its fulfilment and expression. For more on New Age culture, see Hanegraaff, 1998; Wood, 2007; Heelas, 1996; Kaplan & Werczberger, 2017.
18. Illouz & John, 2003.
19. A similar shift—from natural foods stores selling organic food to big organic retailers—is described by sociologist Laura J. Miller in relation to the trajectory of natural food in the United States. See Miller, 2017.

20. For more about Whole Foods Market, a grocery retail chain that operates in North America and the United Kingdom, see Johnston, 2008; Davis, 2017.

21. Beck et al., 2003, p. 8.

22. C. Katz, 2011.

23. Sela, 2012.

24. Johnston, 2008.

25. L. Levin, 2011; Ram, 2016.

26. Pollan, 2007.

27. Johnston, 2008.

28. Shaili, 2006.

29. Yefet, 2009; Sela, 2012; Morganstern, 2011; Kadosh & Shahar-Levi, 2015.

30. Kadosh & Shahar-Levi, 2015.

31. See, for example, Guthman, 2004; Fromartz, 2006; Howard, 2009; Gottschalk & Leistner, 2013; Obach, 2015; Haedicke, 2016; Johnston et al., 2009.

32. Boltanski & Chiapello, 2005. See also Frank, 1998.

33. Block, 1990, p. 21; Granovetter, 1985.

34. Ram, 2016; Ram, 2008; Maron & Shalev, 2017.

35. Dovrat-Mazrich, 2010.

36. Biltekoff, 2013, p. 7.

37. Pollan, 2006.

38. S. Katz & Lavie, 2007.

39. Muzikant, 2003.

40. Ritzer, 2007.

41. Cf. Johnston, 2008.

42. Cf. Johnston et al., 2009.

43. Harvey, 1990.

44. Bourdieu, 1984.

45. Cf. Johnston & Szabo, 2011.

46. Banet-Weiser, 2012, p. 37.

47. Mathews, 2000, p. 18.

48. Johnston & Szabo, 2011.

49. Bauman, 2011, pp. 1–17. See also Campbell, 1987; Veblen, 1899/1934.

50. See similar findings in research on organic food consumption in North America: Johnston & Szabo, 2011; Johnston et al., 2011; Johnston, 2008.

51. Tzahor, 2010.

52. For more about Lavie, see Chapter 6.

53. Lavie, 2008.

54. *Harduf* is oleander in Hebrew.

55. Abramitzky, 2018.

56. Palgi & Reinharz, 2014; Tzfadia, 2008, p. 48.

57. Kibbutz Industries Association website (n.d.).

58. See, for example, Halfin, 2016; Avieli, 2017.

59. Russell et al., 2011; Rosner, 2000; Halfin, 2016; Abramitzky, 2018; Fogiel-Bijaoui, 2007.

60. Getz, 2018.

61. Simchai, 2009, p. 45.

62. Hanegraaff, 1998; Werczberger & Huss, 2014.

63. Lau, 2000.

64. Simchai, 2009; Simchai & Keshet, 2016; Ruah-Midbar & Zaidman, 2013; Werczberger & Huss, 2014; Kaplan & Werczberger, 2017.

65. Werczberger & Huss, 2014. See also Tavory & Goodman, 2009; Klin-Oron, 2014.

66. Beit-Hallahmi, 1992; Ben-Eliezer, 1999; Ram, 2008; Simchai & Keshet, 2016.

67. Anthroposophy emerged as a counterforce within a world of scientific rationalism and evolved as a countercultural lifestyle underpinning the twentieth-century ethos of modernist progress. Rudolf Steiner, who is known as the father of anthroposophy, addressed critically—both theoretically and practically—Enlightenment reasoning, social Darwinism, and the global consequences of the exports of Western civilization. Steiner was also attentive to the empirical knowledge accumulated by European peasantries, especially with reference to indigenous medicinal and agricultural practices deemed backwards, and developed his own biodynamic agricultural ideas. As a spiritual movement, anthroposophy is viewed by many commentators as an important forerunner of the New Age movement. See more in Wood and Bunn, 2009; Brendbekken, 2002; McKanan, 2018.

68. Khamaissi, 2003; Tzfadia, 2008b; H. Levin, 1983.

69. Pinto, 2007.

70. Abramitzky, 2018.

71. See also L. J. Miller, 2017.

72. Simchai, 2009, p. 45.

73. See, for example, Leviatan et al., 1998; Palgi & Reinharz, 2011; Shapira, 2012, p. 386–387.

74. Avieli, 2017.

75. Raviv, 2015, p. 182.

76. The kibbutz dining hall was previously one of the most prominent icons and hallmarks of the Kibbutz Movement. During the time when kibbutzim pursued the ideological logic of communal ownership of property and modes of production and consumption, the "kibbutz dining hall" functioned as a social and cultural hub for kibbutz life (and was even conceived as a "secular synagogue" and "the kibbutz temple"). See Helman, 2014; Ran-Shachnai, 2015; Avieli, 2017.

77. Harstein, 2012, pp. 1–2.

78. Wood, 2007; Heelas, 1996; 2008; Redden, 2011.

79. Katzen, 1977.

80. L. J. Miller & Hardman, 2015, p. 120.

81. Harstein, 2012, p. 15.

82. Harstein, 2012, p. 257.

83. Harstein, 2012, p. 258.

84. Harstein, 2012, p. 16.

85. For more about the notion "spiritual capital," see Verter, 2003.

86. Leschziner, 2015.

87. Werczberger & Huss, 2014.

88. Ram, 2008, pp. 63–64; Fogiel-Bijaui, 2007; Abramitzky, 2018.

89. Kopytoff, 1986.

90. Kimmerling, 2001; Ram, 2008, p. 90; Shafir & Peled, 2002.

91. Fogiel-Bijaui 2007; Halfin, 2016; Getz, 2018.

92. Ram, 1999; 2008, Shalev, 2000.

93. Pinto, 2007.

94. See Kalberg, 1980.

95. Haedicke, 2016.

96. Obach, 2015.

97. Ram, 2008, p. 77.

98. A. Cohen, 2006.

99. Verlinsky, 1960.

100. Rosenthal & Eiges, 2014.

101. Hershkovitz, 2017.

102. Sade, 2011.

103. Bourdieu, 1993; 1993b; Hebdige, 1979; Frank, 1998.

104. See Tnuva, n.d.

105. Lavie, 2008.

106. As of 2020, 32 Waldorf (or Waldorf-inspired) schools and 117 Waldorf (or Waldorf-inspired) preschools operate in Israel. See Anthroposophy in Israel, 2020; 2020b.

107. Howes, 1996.

108. Simchai, 2009, p. 90.

109. Simchai, 2009, pp. 92–97; Simchai & Keshet, 2016.

110. Bruce, 1996.

111. Guthman, 2019; see also Buck et al., 1997.

## Chapter 5

1. Lamont, 1992; Swidler, 1986.

2. Bourdieu, 1984, p. 225.

3. Eco-habitus refers to a set of dispositions and guiding paradigms for sustainable lifestyle choices. See Carfagna et al., 2014. For more about LOHAS, see Emerich, 2011.

4. Matthews & Maguire, 2014, p. 1.

5. Bourdieu, 1984, p. 3; see also pp. 225–256.

6. Lash & Urry, 1987.

7. Nixon & Gay, 2002.

8. Callon et al., 2002, p. 206.

9. Appadurai, 1996.

10. Maguire & Lim, 2015.

11. Frenkel, 2005.

12. Callon & Latour, 1981. See also Pagis et al., 2018; Carlile, 2004; G. S. Drori et al., 2014; Levitt et al., 2013.

13. Goodman, 2013.

14. Lavie, 2004.

15. Boykoff, et al., 2009. See also Goodman, 2010; Goodman & Littler, 2013.

16. Lavie, 2004.

17. Interview with the author.

18. Koren, 2009; Lewis, 2008; 2013.

19. Wolinitz, 1998.

20. Ram, 2008.

21. Wolinitz, 1999.

22. Maabarot Products was originally founded in 1963 by Kibbutz Maabarot to produce milk substitutes for calves, and then it developed the popular dog food brand Bonzo in the 1970s and baby formulas and powders for the food industry under the brand name Materna in the 1980s; it then purchased the Altman and Ta'am Teva companies and marketed vitamin, mineral, and nutritional supplements. In the 2000s, it purchased Adama and entered the field of organic food and later became partners with Osem-Nestlé.

23. Johnston & Cairns, 2012.

24. Soffer, 2015, p. 60.

25. Hollowell, 1977.

26. Soffer, 2015.

27. Regev & Seroussi, 2004.

28. The contribution of the mekomonim to the creation of urban culinary culture in Israel is reflected in the project of Taam Hair ("the taste of the city"), an annual food festival held since 1996 under the auspices of the local newspaper Hair whose goal is "to bring gourmet food to the general public at an affordable price," as is indicated each year in the advertising leaflets for the event.

29. Founded in 1918, Haaretz is the longest-running daily newspaper currently in print in Israel. It is known for its left-wing and liberal stance and it projects a high-quality and elitist image (Soffer, 2015, p. 48).

30. S. Katz & Lavie, 2007b.

31. O. Almog, 2004, p. 34.

32. O. Almog, 2001. For more about the development of a "cult of parenthood" in the United States, see C. C. Miller, 2018.

33. Zelizer, 1985; see also Cairns et al., 2013.

34. Burman & Stacey, 2010, p. 229.

35. Cairns et al., 2013, pp. 100–101; Cairns et al., 2014.

36. See more about the Orbanic Market in Chapter 3.

37. Lavie, 2010.

38. S. Katz & Lavie, 2006.

39. Cf. Szasz, 2007.

40. Dickinson & Carsky, 2005; Devinney et al., 2010.

41. For more on green economy, see Alkon, 2012.

42. Alkon, 2012, p. 22; see also Peck & Tickell, 2002; Guthman, 2008; Harvey, 2005.

43. R. Cohen, 2007.

44. For more about the social history of "eating right" in the context of United States, see Biltekoff, 2013.

45. Talshir, 2012, p. 8.

46. Talshir, 2012.

47. Named after the Kingston Clinic established in 1938 in Edinburgh, Scotland, by James Charles Thomson.

48. Named after Ann Wigmore, a holistic health practitioner and one of the first to popularize ideas about raw food in the US.

49. Named after Victoria Boutenko, best-selling author and raw food advocate, also known as "the mother of all green smoothies."

50. Talshir, 2012, p. 70.

51. Scrinis, 2013.

52. Scrinis, 2008.

53. *Media frame* refers to a central organizing idea for making sense of relevant issues and "organizing the world," both for the journalists who report it and for those who rely on their reports. See Gamson & Modigliani, 1989; Gitlin, 1980.

54. Kaminer, 2019.

55. A genre epitomized in Marion Nestle's best-selling book, *What to Eat* (2006). See also Pollan, 2006; 2013.

56. Guthman, 2007b.

57. Talshir, 2012, p. 8.

58. Vered et al., 2010.

59. Talshir, 2012.

60. O. Almog, 1999.

61. Peterson & Kern, 1996; Johnston & Baumann, 2009.

62. Gur, 2007.

63. For more about the ways in which gastronomic writings serve as foundations for the development of gastronomic fields, see Ferguson, 1998; Fantasia, 2010.

64. Cairns et al., 2010; Finn, 2017.

65. Johnston & Baumann, 2007, p. 179.

66. Johnston & Bauman, 2009; Johnston et al., 2011.

67. Vered, 2012.

68. Vered, 2010.

69. Talshir, 2012.

70. For more about the term "culinary tourism," see Long, 2004.

71. "Food with a face" is the catchphrase that constitutes the gastronomic authenticity of food by establishing it as having an idiosyncratic connection to a specified creative talent or family tradition, thereby distinguishing the food as "quality," artful food. See Johnston & Baumann, 2007.

72. Vered, 2013.

73. Tovey, 1999; L. J. Miller, 2017, p. 206; Guthman, 2019.

74. Hofer, 2000; Padel et al., 2007; Goodman et al., 2012, p. 153; Guthman, 2019; Boström & Klintman, 2006; Obach, 2015, p. 24; Guthman, 2004, pp. 169–171; Haedicke, 2016, p. 174. For a similar analysis regarding Fair Trade labeling, see Jaffee, 2012.

75. Pagis et al., 2018.

76. A Knesset lobby is a group of Knesset members who want to enlist support for a specific topic among their colleagues and/or other government officials.

77. Taken from Economic Affairs Knesset Committee, *Protocol 139*, June 14, 2000.

78. Economic Affairs Knesset Committee, *Protocol 194*, May 17, 2004.

79. Talshir, 2012.

80. Knesset General Meeting, *Protocol 157*, Nov. 22, 2000.

81. Guthman, 2007; 2019. For more about the notion "economization," see Brown, 2015.

82. Vos, 2000; Boström & Klintman, 2006.

83. Economic Affairs Knesset Committee, *Protocol 139*, June 14, 2000.

84. Talshir, 2012.

85. Ram, 2008.

86. Economic Affairs Knesset Committee, *Protocol 139*, June 14, 2000.

87. Economic Affairs Knesset Committee, *Protocol 194*, May 17, 2004.

88. Szasz, 2007.

89. The Hebrew word *freier* (to be a *freier*) might be translated as dupe, as someone who is gullible, or as someone who is easily tricked. Conversely, "not to be a *freier*" could be translated as being no one's fool. For extensive discussion of the source of the concept, see Roniger & Feige, 1992.

90. Alkon & Guthman, 2017, p. 12; see also Guthman, 2011.

91. Lis, 2018.

92. Szasz, 2007; Johnston & Baumann, 2009; L. J. Miller, 2017; Guthman, 2011.

93. Appadurai, 2013. See also Gibson-Graham, 1996/2006.

# Conclusion

1. Ram, 2004; Illouz & John, 2003; Azaryahu, 1999.

2. See, for example, J. Watson, 2006; J. Watson & Caldwell, 2005.

3. Bittman et al., 2017.

4. As mentioned in Chapter 5, *freier* partakes in a vernacular cultural frame and is slang for a sucker or gullible individual; see Roniger & Feige, 1992.

5. Namely, ideas that oppose the destructive effects of economic globalization on environmental and climate protection, human health, and social justice. See Pleyers, 2010.

6. See, for example, Alkon & Guthman, 2017; Alkon & Agyeman, 2011; Guthman, 2008.

7. See, for example, Guthman, 2004; Aistara, 2018; Goodman et al., 2012.

8. Buck et al., 1997; Guthman, 2019.

9. For more about this multilevel model of the relationship between the global and the local, see Ram, 2004.

10. For more about the distinctive definitions of "global cuisine" and "counter cuisine," see Ashley et al., 2004, pp. 91–104; Belasco, 1989/2007.

11. Johnston & Baumann, 2009; Johnston et al., 2011.

12. For more about veganism in Israel, see Weiss, 2016; O. Schwarz, 2020.

13. Bourdieu, 1993b, pp. 72–77.

14. Gutkowski, 2010.

15. Rudy, 2012.

16. Kalai, 2017.

17. See, for example, Mayer-Tzizik, 2012.

18. A similar argument, in relation to food justice movements, can be found in Alkon & Agyeman, 2011, p. 11.

19. Guthman, 2014, p. 207.

20. Adler, 2005, p. 62.

21. Guthman, 2014, p. 218.

22. Bourdieu, 1983; 1996.

23. Calhoun, 2002; Sklair, 2001.

24. Including, but not limited to, eco-habitus (Carfagna et al., 2014) and therapeutic habitus (Illouz, 2008). For examples of eco-habitus, see Chapter 3, as well as Wolinitz, Katz, and Lavie in Chapter 5. For examples of therapeutic habitus, see Talshir in Chapter 5.

25. DuPuis, 2015; Biltekoff, 2013.

26. Milton, 1996.

27. Jørgensen, 2016; Illouz, 2007.

28. Johnston & Baumann, 2009.

29. Johnston, 2008.

30. Regev, 2007.

31. Regev, 2007.

32. For more about the term *doxa*, which refers to the taken-for-granted aspects of any particular society or social field, see Bourdieu, 1984, p. 471; Bourdieu, 1972/1977.

33. Tomlinson, 1999, p. 202; Regev, 2007.

34. DeSoucey, 2016; Guthman, 2008.

35. Johnston, 2008.

36. Johnston, 2008; see also L. J. Miller, 2006, pp. 197–229.

37. DeSoucey, 2016; L. J. Miller, 2017; Hirsch, 2011.

38. The term *food citizenship* is defined as the practice of engaging in food-related behaviors that support, rather than threaten, the development of a democratic, socially and economically just, and environmentally sustainable food system. See Wilkins, 2005; Lockie, 2009; Lyson, 2012.

39. DeSoucey, 2016, p. 206.

40. Guthman, 2011b.

# References

Abramitzky, R. (2018). *The mystery of the Kibbutz: Egalitarian principles in a capitalist world*. Princeton University Press.

Abufarha, N. (2008). Land of symbols: Cactus, poppies, orange and olive trees in Palestine. *Identities: Global Studies in Culture and Power, 15*(3), 343–368.

Abu-Sada, C. (2009). Cultivating dependence: Palestinian agriculture under the Israeli occupation. In Ophir, A., Givoni, M., & Ḥanafī, S. (Eds.), *The power of inclusive exclusion: anatomy of Israeli rule in the occupied Palestinian territories* (pp. 413–433). Zone Books.

Adler, U. (2005). *Introduction to organic agriculture. Textbook: Basic organic agriculture course*. IBOAA. [In Hebrew].

Aistara, G. A. (2018). *Organic sovereignties: Struggles over farming in an age of free trade*. University of Washington Press.

Algazi, G. (2009). Matrix in Bil'in: Colonial capitalism in the occupied territories. In Ophir, A., Givoni, M., & Ḥanafī, S. (Eds.), *The power of inclusive exclusion: Anatomy of Israeli rule in the occupied Palestinian territories* (pp. 519–533). Zone Books.

Alkon, A. H. (2008). Paradise or pavement: The social constructions of the environment in two urban farmers markets and their implications for environmental justice and sustainability. *Local Environment: The Journal of Justice and Sustainability, 13*(3), 271–289.

Alkon, A. H. (2012). *Black, white, and green: Farmers markets, race, and the green economy*. University of Georgia Press.

Alkon, A. H. (2013). The socio-nature of local organic food. *Antipode, 45*(3), 663–680.

Alkon, A. H., & Agyeman, J. (2011). *Cultivating food justice: Race, class, and sustainability*. MIT Press.

Alkon, A. H., & Guthman, J. (2017). *The new food activism: Opposition, cooperation, and collective action*. University of California Press.

Allegra, M. (2013). The politics of suburbia: Israel's settlement policy and the production of space in the metropolitan area of Jerusalem. *Environment and Planning A, 45*(3), 497–516.

Allegra, M., Handel, A., & Maggor, E. (Eds.). (2017). *Normalizing occupation: The politics of everyday life in the West Bank settlements.* Indiana University Press.

Almog, O. (1999). From vegetable salad and cultured milk to hamburger and sushi: The Cocacolonization of Israel. *Makom Le Machshava, 2,* 7–19. [In Hebrew].

Almog, O. (2001). Shifting the centre from nation to individual and universe: The new 'democratic faith' of Israel. *Israel Affairs, 8*(1–2), 31–42.

Almog, O. (2004). One middle class, three different lifestyles: The Israeli case. *Geography Research Forum, 24,* 37–57.

Almog, S. (2000). The metaphorical pioneer vs. age-old diaspora. In Shapira., A, Reinharz, J. & Harris., J. (Eds.) *The age of Zionism* (pp. 91–108). Zalman Shazar Press. [In Hebrew].

Anthroposophy in Israel. (2020). List of kindergartens (Waldorf kindergartens). http://www.antro.co.il/study/kindergartens.html [In Hebrew].

Anthroposophy in Israel. (2020b). List of anthroposophical schools. http://www.antro.co.il/study/schools.html [In Hebrew].

Appadurai, A. (1996). *Modernity at large: Cultural dimensions of globalization.* University of Minnesota Press.

Appadurai, A. (2013). *The future as cultural fact.* Verso.

Appadurai, A. (Ed.) (1986). *The social life of things: Commodities in cultural perspective.* Cambridge University Press.

Arad, D. (2012, April 24). Farming in the West Bank: organic paradise, Thorny reality. *Haaretz.* https://www.haaretz.com/1.5216745

Ariel, A. (2012). The hummus wars. *Gastronomica: The Journal of Food and Culture, 12*(1), 34–42.

Ashley, B., Hollows, J., Steve, J., & Taylor, B. (2004). *Food and cultural studies.* Routledge.

Autumn in the farmers market. (2011, November). *Al Hashulchan,* p. 4.

Avieli, N. (2016). The hummus wars revisited: Israeli-Arab food politics and gastromediation. *Gastronomica: The Journal of Critical Food Studies, 16*(3), 19–30.

Avieli, N. (2017). *Food and power: A culinary ethnography of Israel.* University of California Press.

Azaryahu, M. (1999). McDonald's or Golani Junction? A case of a contested place in Israel. *The Professional Geographer, 51*(4), 481–492.

Banet-Weiser, S. (2012). *Authentic™: The politics of ambivalence in a brand culture.* NYU Press.

Barham, E. (1997). Social movements for sustainable agriculture in France: A Polanyian perspective. *Society & Natural Resources, 10*(3), 239–249.

Barton, G. A. (2018). *The global history of organic farming.* Oxford University Press.

Bar-Zuri, R., & Warshevski, N. (2010). *Sustainable consumption Israel—Report.* Ministry of Economy, Israel. [In Hebrew].

Baudrillard, J. (1994). *Simulacra and simulation.* University of Michigan Press.

Bauman, Z. (1998). *Globalization: The human consequences.* Polity.

Bauman, Z. (2011). *Culture in a liquid modern world.* Polity.

Baumann, S., Engman, A., Huddart-Kennedy, E., & Johnston, J. (2017). Organic vs. local: Comparing individualist and collectivist motivations for "ethical" food consumption. *Canadian Food Studies, 4*(1), 68–86.

Beamish, T. D. (2015). *Community at risk: biodefense and the collective search for security.* Stanford University Press.

Beck, U. (1992). *Risk society: Towards a new modernity.* Sage.

Beck, U., Sznaider, N., & Winter, R. (Eds.) (2003). *Global America? The cultural consequences of globalization.* Liverpool University Press.

Bégin, C. (2016). *Taste of the nation: The New Deal search for America's food.* University of Illinois Press.

Beit-Hallahmi, B. (1992). *Despair and deliverance: Private salvation in contemporary Israel.* State University of New York Press.

Belasco, W. (2007). *Appetite for change: How the counterculture took on the food industry.* Cornell University Press. (Original work published 1989)

Ben-Eliezer, U. (1999). Is civil society emerging in Israel? Politics and identity in the new associations. *Israeli Sociology, 2*(1): 51–98. [In Hebrew].

Berger, T. (2019). Suburban Realities: The Israeli Case. *CLCWeb: Comparative Literature and Culture, 21*(2), article 3.

Beyond the Lines. (2018). Har-Sinai, spiritual organic farm. https://outline.org.il/blog/2018/07/03/חקלאות-אורגנית/ [In Hebrew].

Biltekoff, C. (2013). *Eating right in America: The cultural politics of food and health.* Duke University Press.

Bio Tour. (n.d.). Experience agriculture, explore nature. https://www.bio-tour.com/english/

Bittman, M., Pollan, M., De-Shutter, O., & Salvador, R. (2017, January 16). Food and more: Expanding the movement for the Trump era. *Civil Eats.* https://civileats.com/2017/01/16/food-and-more-expanding-the-movement-for-the-trump-era/

Bivar, V. (2018). *Organic resistance: The struggle over industrial farming in postwar France.* University of North Carolina Press.

Bloch-Tzemach, D. (1998). *Tourist, "dwelling-tourist" and all the rest: The dwelling experience in Japan as a study-case for conceptualizing a new type of tourist* [MA thesis]. Hebrew University of Jerusalem. [In Hebrew].

Block, F. (1990). *Postindustrial possibilities: A critique of economic discourse.* University of California Press.

Boltanski, L., & Chiapello, E. (2005). *The new spirit of capitalism.* Verso.

Bonacich, E. (1972). A theory of ethnic antagonism: The split labor market. *American Sociological Review, 37*(5), 547–559.

Boström, M., & Klintman, M. (2006). State-centered versus nonstate-driven organic food standardization: A comparison of the US and Sweden. *Agriculture and Human Values, 23*(2), 163–180.

Bourdieu, P. (1969). Intellectual field and creative project. *Information Social Science Information*, 8(2), 89–119.

Bourdieu, P. (1977). *Outline of a theory of practice*. Cambridge University Press. (Original work published 1972)

Bourdieu, P. (1983). The field of cultural production, or: The economic world reversed. *Poetics*, 12(4–5), 311–356.

Bourdieu, P. (1984). *Distinction: A social critique of the judgement of taste*. Harvard University Press.

Bourdieu, P. (1989). Social space and symbolic power. *Sociological Theory*, 7(1), 14–25.

Bourdieu, P. (1990). *In other words: Essays towards a reflexive sociology*. Polity.

Bourdieu, P. (1993). *The field of cultural production: Essays on art and literature*. Columbia University Press.

Bourdieu, P. (1993b). *Sociology in question*. Sage. (Original work published 1984)

Bourdieu, P. (1996). *The rules of art: Genesis and structure of the literary field*. Stanford University Press.

Boykoff, M. T., Goodman, M. K., & Curtis, I. (2009). Cultural politics of climate change: Interactions in everyday spaces. In M. Boykoff (Ed.). *The politics of climate change* (pp. 136–154). Routledge.

Brendbekken, M. (2002). Beyond vodou and anthroposophy in the Dominican-Haitian borderlands. *Social Analysis*, 46(3), 31–74.

Brener, T. (2018, August 2). Givo't Olam: A wild organic farm. *E-food: An economic newspaper for decision makers in the food and beverage industry*. [In Hebrew].

Brenner, M. (2018). *In search of Israel: The history of an idea*. Princeton University Press.

Brooks, D. (2000). *Bobos in paradise: The new upper class and how they got there*. Simon & Schuster.

Brown, W. (2015). *Undoing the demos: Neoliberalism's stealth revolution*. Zone Books.

Bruce, S. (1996). *Religion in the modern world: From cathedrals to cults*. Oxford University Press.

Buck, D., Getz, C., & Guthman, J. (1997). From farm to table: The organic vegetable commodity chain of Northern California. *Sociologia Ruralis*, 37(1), 3–20.

Burman, E., & Stacey, J. (2010). The child and childhood in feminist theory. *Feminist Theory* 11(3): 227–224.

Cairns, K., Johnston, J., & Baumann, S. (2010). Caring about food: Doing gender in the foodie kitchen. *Gender & Society*, 24(5), 591–615.

Cairns, K., Johnston, J., & MacKendrick, N. (2013). Feeding the 'organic child': Mothering through ethical consumption. *Journal of Consumer Culture*, 13(2), 97–118.

Cairns, K., de Laat, K., Johnston, J., & Baumann, S. (2014). The caring, committed eco-mom: consumption ideals and lived realities of Toronto mothers.

In Barendregt, B., & Jaffe, R. (Eds.), *Green consumption: The global rise of eco-chic* (pp. 100–114). Bloomsbury.

Caldwell, M. L. (2006). Tasting the worlds of yesterday and today: culinary tourism and nostalgia foods in post-Soviet Russia. In Wilk, R. (Ed.), *Fast food/slow food: The cultural economy of the global food system* (pp. 97–112). Rowman and Littlefield.

Calhoun, C. J. (2002). The class consciousness of frequent travelers: Toward a critique of actually existing cosmopolitanism. *The South Atlantic Quarterly*, 101(4), 869–897.

Callon, M., & Latour, B. (1981). Unscrewing the big leviathan: How actors macro-structure reality and how sociologists help them to do so. In Knorr-Cetina, K., & Cicourel, A. V. (Eds.), *Advances in social theory and methodology: Toward an integration of micro-and macro-sociologies* (pp. 277–303). Routledge.

Callon, M., Méadel, C., & Rabeharisoa, V. (2002). The economy of qualities. *Economy and Society*, 31(2), 194–217.

Campbell, C. (1987). *The romantic ethic and the spirit of modern consumerism*. Blackwell.

Carfagna, L. B., Dubois, E. A., Fitzmaurice, C., Ouimette, M. Y., Schor, J. B., Willis, M., & Laidley, T. (2014). An emerging eco-habitus: The reconfiguration of high cultural capital practices among ethical consumers. *Journal of Consumer Culture*, 14(2), 158–178.

Carlile, P. R. (2004). Transferring, translating, and transforming: An integrative framework for managing knowledge across boundaries. *Organization Science*, 15(5), 555–568.

Carson, R. (1962). *Silent spring*. Houghton Mifflin.

Chapple-Sokol, S. (2013). Culinary diplomacy: Breaking bread to win hearts and minds. *Hague Journal of Diplomacy*, 8(2), 161–183.

Chevannes, B. (1994). *Rastafari: Roots and ideology*. Syracuse University Press.

Clifton, N. (2008). The "creative class" in the UK: An initial analysis. *Geografiska Annaler: Series B, Human Geography*, 90(1), 63–82.

Cloud, J. (2007, March 2). Eating better than organic. *Time*.

Cohen, A. (2006, September 7). Badatz HaEidah HaCharedit canceled Kosher certification for organic products of Harduf. *Haaretz*. https://www.haaretz.co.il/misc/1.1135005 [In Hebrew].

Cohen, E. (2003). Backpacking: Diversity and change. *Journal of Tourism and Cultural Change*, 1(2), 95–110.

Cohen, R. (2007, April 4). Organic propaganda in Haaretz. *News1*. https://www.news1.co.il/Archive/003-D-21316-00.html [In Hebrew].

Cohen, U., & Leon, N. (2008). The new Mizrahi middle class: Ethnic mobility and class integration in Israel. *Journal of Israeli History*, 27(1), 51–64.

Coles, B., & Crang, P. (2011) Placing alternative consumption: Commodity fetishism in Borough Fine Foods Market, London. In Lewis, T., & Potter, E. (Eds.), *Ethical consumption: A critical introduction* (pp. 87–102). Routledge.

Cone, C. & Myhre, A. (2000). Community-supported agriculture: A sustainable alternative to industrial agriculture? *Human Organization*, 59(2), 187–197.

Conford, P. (2001). *The origins of the organic movement*. Floris Books.

Dagan-Buzaglo, N., & Konor-Atias, E. (2013). *Middle class in Israel 1992–2010*. Adva Center. [In Hebrew].

Davis, J. (2017). *From head shops to whole foods: The rise and fall of activist entrepreneurs*. Columbia University Press.

DeSoucey, M. (2016). *Contested tastes: Foie gras and the politics of food*. Princeton University Press.

DeVault, M. (1991). *Feeding the family: The social organization of caring as gendered work*. University of Chicago Press.

Devinney, T., Auger, P., & Eckhardt, G. (2010). *The myth of the ethical consumer*. Cambridge University Press.

De-Vinter, A. (n.d.) Israeli organic agriculture summit. *Kibbutz.org.il*. http://www.kibbutz.org.il/iff/meida/agri_news/060905_mashov2.htm?nojump

Dewey, C. (2017, June 30). The big consequence of the Amazon-Whole Foods deal no one's talking about. *The Washington Post*. https://www.washingtonpost.com/news/wonk/wp/2017/06/30/the-big-consequence-of-the-amazon-whole-foods-deal-no-ones-talking-about/

Dickinson, R. A., & Carsky, M. L. (2005). The consumer as economic voter. In Harrison, R., Newholm, T., & Shaw, D. (Eds.), *The ethical consumer* (pp. 25–36). Sage.

DiMaggio, P. J., & Powell, W. W. (1983). The iron cage revisited: Institutional isomorphism and collective rationality in organizational fields. *American Sociological Review*, 48(2),147–160.

Dixon, J. (1999). A cultural economy model for studying food systems. *Agriculture and Human Values*, 16(2), 151–160.

Douglas, M. (1978). *Purity and danger: An analysis of concepts of pollution and taboo*. Routledge. (Original work published 1966)

Dovrat-Mazrich, A. (2010, July 8). Natural selection: Big food chains go organic. *TheMarker*. https://www.themarker.com/markets/1.575454 [In Hebrew].

Dromi, S. M. (2014). Uneasy settlements: Reparation politics and the meanings of money in the Israeli withdrawal from Gaza. *Sociological Inquiry*, 84(2), 294–315.

Drori, G. S., Höllerer, M. A., & Walgenbach, P. (Eds.). (2014). *Global themes and local variations in organization and management: Perspectives on glocalization*. Routledge.

Drori, M. (2005, June 5). Organic katif. *NRG*. https://www.makorrishon.co.il/nrg/online/1/ART/954/294.html [In Hebrew].

DuPuis, M. (2015). *Dangerous Digestion: The Politics of American Dietary Advice*. University of California Press.

Duram, L. A. (Ed.). (2010). *Encyclopedia of organic, sustainable, and local food.* Greenwood.

Duruz, J. (2004). Adventuring and belonging: An appetite for markets. *Space and Culture, 7*(4), 427–445.

Economic Affairs Knesset Committee. (2000, June 14). *Protocol 139.*

Economic Affairs Knesset Committee. (2004, May 17). *Protocol 194.*

Eden, S. (2011). The politics of certification: Consumer knowledge, power, and global governance in ecolabeling. In Peet, R., Robbins, P., & Watts, M. (Eds.), *Global political ecology* (pp. 169–184). Routledge.

Eisenstadt, S. N. (1967). *Israeli society.* Basic Books.

Eldad, K. & Bashan, S. (2011). *YESHA is Fun.* Self Publication.

Emerich, M. (2011). *The Gospel of sustainability: Media, market, and LOHAS.* University of Illinois Press.

Etkes, D. (2013). *Israeli settler agriculture as a means of land takeover in the West Bank.* Kerem Navot Report. [In Hebrew].

Even-Zohar, I. (1981). The emergence of a native Hebrew culture in Palestine: 1882–1948. *Studies in Zionism, 2*(2), 167–184.

Fantasia, R. (2010). 'Cooking the books' of the French gastronomic field. In Silva, E., & Warde, A. (Eds.), *Cultural analysis and Bourdieu's legacy: Settling accounts and developing alternatives* (pp. 28–44). Routledge.

Feige, M. (2009). *Settling in the hearts: Jewish fundamentalism in the occupied territories.* Wayne State University Press.

Ferguson, P. P. (1998). A cultural field in the making: Gastronomy in 19th-century France. *American Journal of Sociology, 104*(3), 597–641.

Ferguson, P. P. (2011). The Senses of Taste. *The American historical review, 116*(2), 371–384.

Fine, G. A., & DeSoucey, M. (2005). Joking cultures: Humor themes as social regulation in group life. *Humor: International Journal of Humor Research, 18*(1), 1–22.

Fine, G. A., & Wood, C. (2010). Accounting for jokes: Jocular performance in a critical age. *Western Folklore, 69*(3–4), 299–321.

Finkelshtain, I., & Kachel, Y. (2006). *The organization of agricultural exports: Lessons from reforms in Israel.* Discussion Paper No. 6.06. Hebrew University of Jerusalem.

Finn, S. (2017). *Discriminating taste: How class anxiety created the American food revolution.* Rutgers University Press.

Fischer, S. (2011). Radical religious Zionism from the collective to the Individual. In Huss, B. (Ed.). *Kabbalah and contemporary spiritual revival* (pp. 285–309). Ben-Gurion University of the Negev.

Fischler, C. (1979). Gastro-nomie et gastro-anomie. *Communications, 31*(1), 189–210.

Fligstein, N., & McAdam, D. (2012). *A theory of fields.* Oxford University Press.

Florida, R. (2002). *The rise of the creative class: And how it's transforming work, leisure, community and everyday life*. Basic Books.

Florida, R. (2004). America's looming creativity crisis. *Harvard Business Review*, 82 (10), 122–136.

Fogiel-Bijaoui, S. (2007). Women in the kibbutz: The "mixed blessing" of neo-liberalism. *Nashim: A Journal of Jewish Women's Studies & Gender Issues, 13*, 102–122.

Frank. T. (1998) *The conquest of cool: Business culture, counterculture, and the rise of hip consumerism*. University of Chicago Press.

Frenkel, M. (2005). The politics of translation: How state-level political relations affect the cross-national travel of management ideas. *Organization, 12*(2), 275–301.

Fromartz, S. (2006) *Organic, Inc.: Natural foods and how they grew*. Harcourt.

Galt, R. E. (2013). The moral economy is a double-edged sword: Explaining farmers' earnings and self-exploitation in community-supported agriculture. *Economic Geography, 89*(4), 341–365.

Gamson, W. A., & Modigliani, A. (1989). Media discourse and public opinion on nuclear power: A constructionist approach. *American Journal of Sociology, 95*(1), 1–37.

Garía-Álvarez, E., Katz-Gerro, T., & López-Sintas, J. (2007). Deconstructing cultural omnivorousness 1982–2002: heterology in Americans' musical preferences. *Social Forces, 86*(2), 417–443.

Geier, B. (2007). IFOAM and the history of the international organic movement. In Lockeretz, W. (Ed.), *Organic Farming: An International History* (pp. 175–186). CABI.

Getz, S. (2018). Cooperative kibbutzim—Status quo for the beginning of 2018. Institute of Kibbutz Research, Haifa University. http://kibbutz.haifa.ac.il/images/publications/shitufi2017.pdf [In Hebrew].

Gibson-Graham, J. K. (2006). *The end of capitalism (as we knew it): A feminist critique of political economy*. University of Minnesota Press. (Original work published in 1996)

Giddens, A. (1990). *The consequences of modernity*. Stanford University Press.

Gieryn, T. F. (1983). Boundary-work and the demarcation of science from non-science: Strains and interests in professional ideologies of scientists. *American Sociological Review, 48*(6), 781–795.

Gill, R., & Pratt, A. (2008). In the social factory? Immaterial labour, precariousness and cultural work. *Theory, Culture & Society, 25*(7–8), 1–30.

Gitlin, T. (1980). *The whole world is watching: Mass media in the making and unmaking of the New Left*. University of California Press.

Goffman, E. (1959). *The presentation of self in everyday life*. Anchor Books.

Goldberg, A., Hannan, M. T., & Kovács, B. (2016). What does it mean to span cultural boundaries? Variety and atypicality in cultural consumption. *American Sociological Review, 81*(2), 215–241.

Goodman, D. (2000). Organic and conventional agriculture: Materializing discourse and agro-ecological managerialism. *Agriculture and Human Values*, *17*(3), 215.

Goodman, D., DuPuis, E. M., & Goodman, M. K. (2012). *Alternative food networks: Knowledge, practice, and politics*. Routledge.

Goodman, M. K. (2010). The mirror of consumption: Celebritization, developmental consumption and the shifting cultural politics of fair trade. *Geoforum*, *41*(1), 104–116.

Goodman, M. K. (2013). Celebritus politicus, neoliberal sustainabilities and the terrains of care. In Fridell, G., & Konings, M. (Eds.), *Age of icons: Exploring philanthrocapitalism in the contemporary world* (pp. 72–92). University of Toronto Press.

Goodman, M. K., & Littler, J. (2013). Celebrity ecologies: Introduction. *Celebrity Studies*, *4*(3), 269–275.

Gottschalk, I., & Leistner, T. (2013). Consumer reactions to the availability of organic food in discount supermarkets. *International Journal of Consumer Studies*, *37*(2), 136–142.

Gouldner, A. W. (1976). *The dialectic of ideology and technology: The origins, grammar, and future of ideology*. Seabury Press.

Granovetter, M. (1985). Economic action and social structure: The problem of embeddedness. *American Journal of Sociology*, *91*(3), 481–510.

Grosglik, R. (2011). Organic hummus in Israel: Global and local ingredients and images. *Sociological Research Online*, *16*(2), 1–11.

Grosglik, R. (2017). Citizen-consumer revisited: The cultural meanings of organic food consumption in Israel. *Journal of Consumer Culture*, *17*(3), 732–751.

Grosglik, R., & Lerner, J. (2020). Gastro-emotivism: How MasterChef Israel produces therapeutic collective belongings. *European Journal of Cultural Studies*. https://doi.org/10.1177/1367549420902801

Grosglik, R., & Ram, U. (2013). Authentic, speedy and hybrid: Representations of Chinese food and cultural globalization in Israel. *Food, Culture & Society*, *16*(2), 223–243.

Gross, A. (2017). *The writing on the wall: Rethinking the international law of occupation*. Cambridge University Press.

Gross, R. (2019, March 28). A growth of the organic by 20%. *Actoalik*. http://actualic.co.il [In Hebrew].

Gur, J. (2007). *The book of New Israeli food: A culinary journey*. Schocken Books.

Guthman, J. (2003). Fast food/organic food: Reflexive tastes and the making of 'yuppie chow.' *Social & Cultural Geography*, *4*(1), 45–58.

Guthman, J. (2004). *Agrarian dreams: The paradox of organic farming in California*. University of California Press.

Guthman, J. (2004b). The trouble with 'organic lite' in California: a rejoinder to the 'conventionalisation' debate. *Sociologia Ruralis*, *44*(3), 301–316.

Guthman, J. (2007). The Polanyian way? Voluntary food labels as neoliberal governance. *Antipode*, 39(3), 456–478.

Guthman, J. (2007b). Commentary on teaching food: Why I am fed up with Michael Pollan et al. *Agriculture and Human Values*, 24(2), 261–264.

Guthman, J. (2008). Neoliberalism and the making of food politics in California. *Geoforum*, 39(3), 1171–1183.

Guthman, J. (2011). *Weighing in: Obesity, food justice, and the limits of capitalism.* University of California Press.

Guthman, J. (2011b). If they only knew: The unbearable whiteness of alternative food. In Alkon, A. H., & Agyeman, J. (Eds.), *Cultivating food justice: Race, class, and sustainability* (pp. 263–281). MIT Press.

Guthman, J. (2014). *Agrarian dreams: The paradox of organic farming in California.* University of California Press.

Guthman, J. (2016). Going both ways: More chemicals, more organics, and the significance of land in post-methyl bromide fumigation decisions for California's strawberry industry. *Journal of Rural Studies*, 47(A), 76–84.

Guthman, J. (2019). The (continuing) paradox of the organic label. In Phillipov, M., & Kirkwood, K. (Eds.), *Alternative food politics: From the margins to the mainstream.* (pp. 23–36). Routledge.

Gutkowski, N. (2010). *The Green Line and the equator: Local Fair Trade and the olive oil sector* [MA thesis]. Tel Aviv University. [In Hebrew].

Gutkowski, N. (2016). *Timely cultivation: Sustainable agriculture policy, temporality and the Palestinian-Arab citizens of Israel* [PhD dissertation]. Tel Aviv University.

Gutkowski, N. (2018). Governing through timescape: Israeli sustainable agriculture policy and the Palestinian-Arab citizens. *International Journal of Middle East Studies*, 50(3), 471–492.

Gvion, L. (2012). *Beyond hummus and falafel: Social and political aspects of Palestinian food in Israel.* University of California Press.

Gvion, L. (2017). Space, gentrification and traditional open-air markets: How do vendors in the Carmel market in Tel Aviv interpret changes? *Community, Work & Family*, 20(3), 346–365.

Habermas, J. (1970). *Toward a rational society: Student protest, science, and politics.* Beacon Press.

Habermas, J. (1981). Modernity versus postmodernity. *New German Critique* (22), 3–14.

Haedicke, M. A. (2016). *Organizing organic: Conflict and compromise in an emerging market.* Stanford University Press.

Haklai, O. (2007). Religious—nationalist mobilization and state penetration: Lessons from Jewish settlers' activism in Israel and the West Bank. *Comparative Political Studies*, 40(6), 713–739.

Halevi E. (2005, September 23). Avraham (Avri) Ran, the revolutionary father of the hilltop outposts. *Arutz Sheva.* http://www.israelnationalnews.com/News/News.aspx/90208 [In Hebrew].

Halfin, T. (2016). *Communal: Utopian daily life on the kibbutz in the 2000s* [PhD dissertation]. Ben-Gurion University of the Negev. [In Hebrew].

Halpern, B., & Reinharz, J. (1998). *Zionism and the creation of a new society.* Brandeis University Press.

Handel, A., Rand, G., & Allegra, M. (2015). Wine-washing: Colonization, normalization, and the geopolitics of terroir in the West Bank's settlements. *Environment and Planning A, 47*(6), 1351–1367.

Hanegraaff, W. (1998). *New Age religion and Western culture.* Brill.

Hannerz, U. (1996). *Transnational connections: Culture, people, places.* Routledge.

Hardt, M., & Negri, A. (2004). *Multitude: War and democracy in the age of empire.* Penguin Press.

Har-Sinai. (2020). About. https://har-sinai.co.il/about/ [in Hebrew].

Harstein, J. (2012). *The living kitchen: Organic vegetarian cooking for family and friends.* Floris Books.

Harvey, D. (1990). Between space and time: Reflections on the geographical imagination. *Annals of the Association of American Geographers, 80*(3), 418–434.

Harvey, D. (2005). *A brief history of neoliberalism.* Oxford University Press.

Hebdige, D. (1979). *Subculture, the meaning of style.* Methuen.

Heelas, P. (1996). *The New-Age movement: The celebration of the Self and the sacralization of modernity.* Blackwell.

Heelas, P. (2008) *Spiritualities of life: New Age romanticism and consumptive capitalism.* Blackwell Publishing.

Helman, A. (2014). *Becoming Israeli: National ideals and everyday life in the 1950s.* Brandeis University Press.

Hershkovitz, S. (2017). "Not buying cottage cheese": Motivations for consumer protest—the case of the 2011 protest in Israel. *Journal of Consumer Policy, 40*(4), 473–484.

Herskovitz, R. (2011, November). The development of organic market [Press announcement]. Ministry of Agriculture and Rural Development, Israel (MARD). https://www.moag.gov.il/yhidotmisrad/research_economy_strategy/publication/2012/pages/orgmardev.aspx. [In Hebrew].

Herzfeld, M. (2016). *Cultural intimacy: Social poetics in the nation-state.* Routledge.

Hinrichs, C. C. (2000). Embeddedness and local food systems: notes on two types of direct agricultural market. *Journal of Rural Studies, 16*(3), 295–303.

Hinrichs, C. C. (2003). The practice and politics of food system localization. *Journal of Rural Studies, 19*(1), 33–45.

Hirsch, D. (2011). "Hummus is best when it is fresh and made by Arabs": The gourmetization of hummus in Israel and the return of the repressed Arab. *American Ethnologist, 38*(4), 617–630.

Hirsch, D. (2015). Hygiene, dirt and the shaping of a new man among the early Zionist halutzim. *European Journal of Cultural Studies, 18*(3), 300–318.

Hirsch, D. (2016). Hummus masculinity in Israel. *Food, Culture & Society, 19*(2), 337–359.

Hirsch, D., & Tene, O. (2013). Hummus: The making of an Israeli culinary cult. *Journal of Consumer Culture, 13*(1), 25–45.

Hirschhorn, S. Y. (2017). *City on a hilltop: American Jews and the Israeli settler movement.* Harvard University Press.

Hobsbawm, E. (1983). Introduction: Inventing traditions. In Hobsbawm, E., & Ranger, T. (Eds.), *The invention of tradition* (pp. 1–14). Cambridge University Press.

Hofer, K. (2000). Labelling of organic food products. In Mol, A. P., Lauber, V., & Liefferink, D. (Eds.), *The voluntary approach to environmental policy: Joint environmental policy-making in Europe* (pp. 156–191). Oxford University Press.

Holloway, L., & Kneafsey, M. (2000). Reading the space of the farmers' market: A preliminary investigation from the UK. *Sociologia Ruralis, 40*(3), 285–299.

Hollowell, J. (1977). *Fact & fiction: The new journalism and the nonfiction novel.* University of North Carolina Press.

Holmstedt, R. D. (2010). *Ruth: A handbook on the Hebrew text.* Baylor University Press.

Howard, P. H. (2009). Consolidation in the North American organic food processing sector, 1997 to 2007. *International Journal of Sociology of Agriculture & Food, 16*(1), 13–30.

Howard, P. H. (2016). *Concentration and power in the food system: Who controls what we eat?* Bloomsbury.

Howes, D. (Ed.). (1996). Cross-cultural consumption: Global markets, local realities. Routledge.

Ichijo, A., & Ranta, R. (2016). *Food, national identity and nationalism: From everyday to global politics.* Palgrave Macmillan.

IFOAM. (2008). Definition of organic agriculture. https://www.ifoam.bio/en/organic-landmarks/definition-organic-agriculture

IFOAM. (2012). The world of organic agriculture: Statistics and emerging trends—2012. IFOAM and FIBL.

IFOAM. (2018). The world of organic agriculture: Statistics and emerging trends—2018. IFOAM and FIBL.

IFOAM. (n.d.). The four principles of organic agriculture. https://www.ifoam.bio/why-organic/shaping-agriculture/four-principles-organic

IFOAM. (n.d.(b)). Principles of organic agriculture: The principle of fairness. https://www.ifoam.bio/en/principles-organic-agriculture/principle-fairness

Ilbery, B., & Maye, D. (2005). Alternative (shorter) food supply chains and specialist livestock products in the Scottish–English borders. *Environment and Planning A, 37*(5), 823–844.

Illouz, E. (2007). *Cold intimacies: The making of emotional capitalism.* Polity.

Illouz, E. (2008). *Saving the modern soul: Therapy, emotions and the culture of self-help.* University of California Press.

Illouz, E., & John, N. (2003). Global habitus, local stratification, and symbolic struggles over identity: The case of McDonald's Israel. *American Behavioral Scientist, 47*(2), 201–229.

Inbar, M., & Osman, E. (2018). *A review of the egg industry in Israel.* MARD. https://www.moag.gov.il/yhidotmisrad/research_economy_strategy/publication/2018/Documents/eggs_2017.pdf [In Hebrew].

Inglis, D., & Gimlin, D. (Eds.). (2009). *The globalization of food.* Berg.

Israel Port Company. (2009, August 10). Export of organic fruit, vegetables up. http://www.israports.org.il/en/IsraelPortCompany/Pages/News/2009/export%20of%20organic%20fruit,%20vegetables%20up.aspx

Jaffee, D. (2012). Weak coffee: Certification and co-optation in the Fair Trade movement. *Social Problems, 59*(1), 94–116.

Jaffee, D., & Howard, P. (2010). Corporate cooptation of organic and fair-trade standards. *Agriculture and Human Values, 27*(4), 387–399.

Jasanoff, S. (2004). *States of knowledge: The co-production of science and the social order.* Routledge.

Jasanoff, S. (2005). *Designs on nature: Science and democracy in Europe and the United States.* Princeton University Press.

Jasanoff, S. (2012). *Science and public reason.* Routledge.

Johnston, J. (2007). Counter-hegemony or bourgeois piggery? Food politics and the case of FoodShare. In Wright, W., & Middendorf, G. (Eds.), *The fight over food: Producers, consumers and activists challenge the global food system* (pp. 93–119). Pennsylvania State University Press.

Johnston, J. (2008). The citizen-consumer hybrid: Ideological tensions and the case of Whole Foods Market. *Theory and Society, 37*(3), 229–270.

Johnston, J., & Baumann, S. (2007). Democracy versus distinction: A study of omnivorousness in gourmet food writing. *American Journal of Sociology, 113*(1), 165–204.

Johnston, J., & Baumann, S. (2009). *Foodies: Democracy and distinction in the gourmet foodscape.* Routledge.

Johnston, J., Biro, A., & MacKendrick, N. (2009). Lost in the supermarket: The corporate-organic foodscape and the struggle for food democracy. *Antipode, 41*(3), 509–532.

Johnston, J., & Cairns, K. (2012). Eating for change. In Banet-Wiser, S., & Mukherji, R. (Eds.), *Commodity activism: Cultural resistance in neoliberal times* (pp. 219–239). NYU Press.

Johnston, J., & Szabo, M. (2011). Reflexivity and the Whole Foods Market consumer: The lived experience of shopping for change. *Agriculture and Human Values, 28*(3), 303–319.

Johnston, J., Szabo, M., & Rodney, A. (2011). Good food, good people: Understanding the cultural repertoire of ethical eating. *Journal of Consumer Culture, 11*(3), 293–318.

Jordan, J. A. (2007). The heirloom tomato as cultural object: Investigating taste and space. *Sociologia Ruralis, 47*(1), 20–41.

Jordan, J. A. (2015). *Edible memory: The lure of heirloom tomatoes and other forgotten foods.* University of Chicago Press.

Jørgensen, M. B. (2016). Precariat—What it is and isn't—Towards an understanding of what it *does. Critical Sociology, 42*(7–8), 959–974.

Julier, A. P. (2013). *Eating together: Food, friendship and inequality.* University of Illinois Press.

Kadosh, N., & Shahar-Levi, Z. (2015, August 17). Who will lose from the sale of Eden Teva Market to Tiv Taam? *Calcalist.* https://www.calcalist.co.il/marketing/articles/0,7340,L-3667098,00.html [In Hebrew].

Kahal, Y. (2007). *Israel and the organic world market—report.* Ministry of Agriculture and Rural Development, Israel. [In Hebrew].

Kalai, H. (2017). Don't play with food: Food regulation in Israel: Between liberalism and sustainability. In Gross, A., & Tirosh, Y (Eds.), *Law and food* (pp. 159–204). Tel-Aviv University Press. [In Hebrew].

Kalberg, S. (1980). Max Weber's types of rationality: Cornerstones for the analysis of rationalization processes in history. *American Journal of Sociology, 85*(5), 1145–1179.

Kamen, C, S. (1991). *Little common ground: Arab agriculture and Jewish settlement in Palestine, 1920–1948.* University of Pittsburgh Press.

Kaminer, M. (2019). *By the sweat of other brows: Thai migrant labor and the transformation of Israeli settler agriculture* [PhD dissertation]. University of Michigan.

Kaplan, D. (2013). Food and class distinction at Israeli weddings: New middle-class omnivores and the "simple taste." *Food, Culture & Society, 16*(2), 245–264.

Kaplan, D., & Werczberger, R. (2017). Jewish New Age and the middle class: Jewish identity politics in Israel under Neoliberalism. *Sociology, 51*(3), 575–591.

Karish-Hazoni, H. (2016, October 19). The widow of the man who was murdered in the terrorist attack claims: 'If you want to take revenge—that's the way!' *MakorRishon, NRG.* https://www.makorrishon.co.il/nrg/online/1/ART2/841/872.html. [In Hebrew].

Katz, C. (2011, November 15). Eden Teva: Israel's answer to Whole Foods. *Forward.* https://forward.com/food/146243/eden-teva-israel-s-answer-to-whole-foods/

Katz, S., & Lavie, A. (2006, December 9). The green party: Organic section. *Haaretz.* [In Hebrew].

Katz, S., & Lavie, A. (2007, July 24). On the way to Eden. *Haaretz.* https://www.haaretz.co.il/misc/1.1419814 [In Hebrew].

Katz, S., & Lavie, A. (2007b). *The Israeli guide to organic food.* Kineret Zmora-Bitan. [In Hebrew].

Katz-Gerro, T. (2009). New middle class and environmental lifestyle in Israel. In H. Lange, H., & Meier, L. (Eds.), *The new middle classes: Globalizing lifestyles, consumerism and environmental concern* (pp. 197–215). Springer.

Katzen, M. (1977). *The Moosewood cookbook.* Ten Speed Press.

Kaufman, D. (2007, November 4). Instant weekend—Tel Aviv. *The Guardian.* https://www.theguardian.com/travel/2007/nov/04/telaviv.escape

Kellerman, A. (1993). *Society and settlement: Jewish land of Israel in the twentieth century.* State University of New York Press.

Kennedy, D. (1998). Shakespeare and cultural tourism. *Theatre Journal, 50*(2), 175–188.

Khamaissi R. (2003). Mechanism of land control and territorial Judaization in Israel In Al-Haj, M., & Ben-Eliezer, U. (Eds.), *In the name of security: Studies in peace and war in Israel in changing times* (pp. 421–448). University of Haifa Press. [In Hebrew].

Kibbutz Industries Association. (n.d.). Kibbutz Industry Association. http://www.kia.co.il/eng/

Kibbutz Sde Eliyahu. (n.d.). Kibbutz Sde Eliyahu. http://www.sde.org.il/Pages/12/About_Us

Kimmerling, B. (2001). *The invention and decline of Israeliness: State, society, and the military.* University of California Press.

Kimmerling, B., & Migdal, J. (1993). *Palestinians: The making of a people.* Free Press.

*King James Bible.* (2004). TruthBeTold Ministry. (Original work published 1611)

Kislev, Yoav. (2015). Agricultural cooperatives in Israel: Past and present. In Kimhi, A., & Lerman, Z. (Eds.), *Agricultural transition in post-Soviet Europe and central Asia after 25 Years* (pp. 281–302). IAMO Press.

Klin-Oron, A. (2014). How I learned to channel: Epistemology, phenomenology, and practice in a New Age course. *American Ethnologist, 41*(4), 635–647.

Kloppenburg, J., Hendrickson, J., & Stevenson, G. W. (1996). Coming in to the foodshed. *Agriculture and Human Values, 13*(3), 33–42.

Knesset General Meeting. (2000, November 22). *Protocol 157.*

Kopytoff, I. (1986). The cultural biography of things: Commoditization as process. In Appadurai, A. (Ed.), *The social life of things: Commodities in cultural perspective* (pp. 64–94). Cambridge University Press.

Koren, G. (2009, March 15). After a heart attack: The organic revolution of Dalik. *Ynet.* https://www.ynet.co.il/articles/0,7340,L-3670846,00.html [In Hebrew].

Kornet, E. (1993). Organic produce export—summary. *Renewed Agriculture, 23*:8. [In Hebrew].

Kotef, H. (2017, November 16). *The love of violence: On chickens, home, and the geographies of exclusion* [Paper presentation]. PARSE conference, University of Gothenburg.

Krampf, A. (2018). *The Israeli path to neoliberalism: The state, continuity and change*. Routledge.

Lamont, M. (1992). *Money, morals, and manners: The culture of the French and the American upper-middle class*. University of Chicago Press.

Laor, Y. (2017). *West side*. Afik Press.

Lapid, Y. (2010, July 5). A piano in a goat pen. *Yedioth Ahronoth*. http://article.yedioth.co.il/default.aspx?articleid=6907 [In Hebrew].

Lash, S. (1990). *The sociology of postmodernism*. Routledge.

Lash, S., & Urry, J. (1987). *The end of organized capitalism*. University of Wisconsin Press.

Latour, B. (1988). *The pasteurization of France*. Harvard University Press.

Lau, K. J. (2000). *New age capitalism: Making money East of Eden*. University of Pennsylvania Press.

Lavie, A. (2003, April 7). The fear of the hills. *Haaretz*. https://www.haaretz.co.il/1.874468 [In Hebrew].

Lavie, A. (2003b, April 9). The sheriff. *Haaretz*. https://www.haaretz.com/1.4837883

Lavie, A. (2004, February 16). Crossing the green line. *Haaretz*. https://www.haaretz.co.il/misc/1.946941 [In Hebrew].

Lavie, A. (2008, October 17). Organic food for the masses. *NRG Maariv*. [In Hebrew].

Lavie, A. (2009, May 29). Wiping hummus at Abu organic. *NRG*. http://www.nrg.co.il/online/1/ART1/895/872.html [In Hebrew].

Lavie, A. (2010, July 25). *Hatachana*: Come and see what a real market is. *Time Tel-Aviv*. https://www.makorrishon.co.il/nrg/online/54/ART2/136/082.html [In Hebrew].

Leadbeater, C., & Oakley., K (1999). *The independents: Britain's new cultural entrepreneurs*. Demos.

Leitch, A. (2012). Slow Food and the politics of "virtuous globalization." In Counihan, C., & Van Esterik, P. (Eds.), *Food and Culture: A Reader* (3rd ed., pp. 423–439). Routledge.

Leschziner, V. (2015). *At the chef's table: Culinary creativity in elite restaurants*. Stanford University Press.

Levi, M. (1993). Reflections. *Renewed Agriculture*, 22:38. [In Hebrew].

Levi, M. (2016). *Lifetime achievement—Autobiography*. IBOAA. [In Hebrew].

Leviatan, U., Oliver, H., & Quarter, J. (Eds.) (1998). *Crisis in the Israeli Kibbutz: Meeting the challenge of changing times*. Praeger.

Levin, H. (1983, October 7). The little Moshav wants to be a big Moshav. *Davar*, 20.

Levin, L. (2011, November 3). The supermarket is acclimating to Israeli society. *Haaretz*. https://www.haaretz.co.il/1.1556073 [In Hebrew].

Lévi-Strauss, C. (2012). The culinary triangle. In Counihan, C., & Van Esterik, P. (Eds.), *Food and culture: A reader* (3rd ed., pp. 40–47). Routledge. (Original work published 1968)

Levit, M. (2018). *The exotic local: A critical reading into the ways in which local wild food is perceived in Israel, the case of Hubeiza* [MA thesis]. University of Gastronomic Sciences.

Levitt, P., Merry, S. E., & Alayza, R. (2013). Doing vernacularization: The encounter between global and local ideas about women's rights in Peru. In Caglar, G., Prügl, E., & Zwingel, S. (Eds.), *Feminist strategies in international governance* (pp. 127–142). Routledge.

Lewis, T. (2008). Transforming citizens? Green politics and ethical consumption on lifestyle television. *Continuum: Journal of Media & Cultural Studies, 22*(2), 227–240.

Lewis, T. (Ed.). (2013). *TV transformations: Revealing the makeover show.* Routledge.

Liechty, M. (2017). *Far out: Countercultural seekers and the tourist encounter in Nepal.* University of Chicago Press.

Lis, J. (2018, October 9). Under right-wing pressure, Israeli army okays extending two agricultural laws to West Bank. *Haaretz.* https://www.haaretz.com/israel-news/.premium-under-pressure-idf-okays-extending-two-agricultural-laws-to-w-bank-1.6544925

Lloyd, R. (2010). *Neo-Bohemia: Art and commerce in the post-industrial city.* Routledge.

Lockeretz, W. (Ed.) (2007). *Organic farming an international history.* CABI.

Lockie, S. (2009). Responsibility and agency within alternative food networks: Assembling the "citizen consumer." *Agriculture and Human Values, 26*(3), 193–201.

Long, L. M. (Ed.) (2004). *Culinary tourism.* University Press of Kentucky.

Lupton, D. (1996). *Food, the body and the self.* Sage.

Lyson, T. A. (2012). *Civic agriculture: Reconnecting farm, food, and community.* University Press of New England.

MacCannell, D. (1976). *The tourist: A new theory of the leisure class.* Schocken Books.

Maguire, J. S., & Lim, M. (2015). Lafite in China: Media representations of 'wine culture' in new markets. *Journal of Macromarketing, 35*(2), 229–242.

MARD. (2005). *Report, district coordination and liaison administration during the evacuation of Gush Katif.* Ministry of Agriculture and Rural Development, Israel (MARD). https://www.moag.gov.il/yhidotmisrad/teum_kishur/teum_aza/publication/2005/documents/peilut_minhelet_tium_kishur_hitnatkut.doc [In Hebrew].

MARD. (n.d.). A list of organic certified farmers. Ministry of Agriculture and Rural Development, Israel (MARD). https://www.moag.gov.il/ppis/Yechidot/standarts/haklaut_organit/pirsumim/2019/Pages/oskim_klalit.aspx [In Hebrew].

Maron, A., & Shalev, M. (Eds.). (2017). *Neoliberalism as a state project: Changing the political economy of Israel.* Oxford University Press.

Martin, J. L. (2003). What is field theory? *American Journal of Sociology, 109*(1), 1–49.

Marx, K. (1973). *Grundrisse: Foundations of the critique of political economy*. Allen Lane, Penguin Books in association with New Left Review. (Original work published 1857)

Mathews, G. (2000). *Global culture/individual identity: Searching for home in the cultural supermarket*. Routledge.

Matthews, J., & Maguire, J. S. (2014). Introduction: Thinking with cultural intermediaries. In Maguire, J. S., & Matthews, J. (Eds.), *The cultural intermediaries reader* (pp. 1–12). Sage.

Mayer-Tzizik, U. (2012). *The local food book*. Mapa Press. [In Hebrew].

Mazori, D. (2005, March 23). D&B: Organic food consumption in Israel rose by 40% in two years. NRG. http://www.nrg.co.il/online/16/ART/886/920.html [In Hebrew].

McKanan, D. (2018). *Eco-Alchemy: Anthroposophy and the history and future of environmentalism*. University of California Press.

McKee, E. 2016. *Dwelling in conflict: Land, belonging and exclusion in the Negev*. Stanford University Press.

Meneley, A. (2007). Like an extra virgin. *American Anthropologist, 109*(4), 678–687.

Meneley, A. (2011). Blood, sweat and tears in a bottle of Palestinian extra-virgin olive oil. *Food, Culture & Society, 14*(2), 275–292.

Meneley, A. (2014). Discourses of distinction in contemporary Palestinian extra-virgin olive oil production. *Food and Foodways, 22*(1–2), 48–64.

Meshek Achiya. (n.d.). Meshek Achiya olive oil. https://offers.israel365.com/meshek-achiya-olive-oil/

Miller, C. C. (2018, December 25). The relentlessness of modern parenting. *New York Times*. https://www.nytimes.com/2018/12/25/upshot/the-relentlessness-of-modern-parenting.html

Miller, L. J. (2006). *Reluctant capitalists: Bookselling and the culture of consumption*. University of Chicago Press.

Miller, L. J. (2017). *Building nature's market: The business and politics of natural foods*. University of Chicago Press.

Miller, L. J., & Hardman, E. (2015). By the pinch and the pound: Less and more protest in American vegetarian cookbooks from the nineteenth century to the present. In Baughman, J. L., Ratner-Rosenhagen, J., & Danky, J. P. (Eds.), *Protest on the page: Essays on print and the culture of dissents since 1865* (pp. 111–136). University of Wisconsin Press.

Milton, K. (1996). *Environmentalism and cultural theory: Exploring the role of anthropology in environmental discourse*. Routledge.

Monterescu, D. (2017). Border wines: Terroir across contested territory. *Gastronomica: The Journal of Critical Food Studies, 17*(4), 127–140.

Montrie, C. (2018). *The myth of silent spring: Rethinking the origins of American environmentalism*. University of California Press.

Morgan, K., Marsden, T., & Murdoch, J. (2006). *Worlds of food: Place, power, and provenance in the food chain.* Oxford University Press.

Morganstern, R. (2011, March 7). Eden Teva: Suphersal is not a threat. *NRG.* https://www.makorrishon.co.il/nrg/online/16/ART2/219/169.html [In Hebrew].

Morris, B. (2008). *1948: A history of the first Arab-Israeli War.* Yale University Press.

Morris, C., & Buller, H. (2003). The local food sector: A preliminary assessment of its form and impact in Gloucestershire. *British Food Journal, 105*(8), 559–566.

Musikant, R. (2003, December 25). Organic for all: The first natural foods supermarket. *Ynet.* https://www.ynet.co.il/articles/0,7340,L-2847825,00.html [In Hebrew].

Nestle, M. (2006). *What to eat.* North Point Press.

Neumann, B. (2011). *Land and desire in early Zionism.* Brandeis University Press.

Nitzan, J., & Bichler, S. (2009). *Capital as power: A study of order and creorder.* Routledge.

Nixon, S. & Gay, P. D. (2002). Who needs cultural intermediaries? *Cultural Studies, 16*(4), 495–500.

Obach, B. K. (2007). Theoretical interpretations of the growth in organic agriculture: Agricultural modernization or an organic treadmill? *Society & Natural Resources, 20*(3), 229–244.

Obach, B. K. (2015). *Organic struggle: The movement for sustainable agriculture in the United States.* MIT Press.

Ocejo, R. (2017). *Masters of craft: Old jobs in the new urban economy.* Princeton University Press.

Oosterveer, P., & Sonnenfeld, D. A. (2011). *Food, globalization and sustainability.* Routledge.

Ophir, A., Givoni, M., & Ḥanafī, S. (Eds.). (2009). *The power of inclusive exclusion: Anatomy of Israeli rule in the occupied Palestinian territories.* Zone Books.

An organic fraud in Gush Katif. (1997, July 7). *Globes.* http://www.globes.co.il/news/article.aspx?did=676

Orlando, G. (2018). Offsetting risk: Organic food, pollution, and the transgression of spatial boundaries. *Culture, Agriculture, Food and Environment, 40*(1), 45–54.

Ortner, S. B. (1973). On key symbols. *American Anthropologist, 75*(5), 1338–1346.

Padel, S., Lampkin, N., & Lockeretz, W. (2007). The development of governmental support for organic farming in Europe. In Lockeretz, W. (Ed.). *Organic farming: an international history* (pp. 93–122). CABI.

Pagis, M., Cadge, W., & Tal, O. (2018). Translating spirituality: Universalism and particularism in the diffusion of spiritual care from the United States to Israel. *Sociological Forum, 33*(3), 596–618.

Palgi, M., & Reinharz, S. (Eds.). (2014). *One hundred years of kibbutz life: A century of crises and reinvention.* Transaction.

Palti, M. (2007, January 2). The eggs from the hills. *Haaretz*. https://www.haaretz. co.il/gallery/1.1376117 [In Hebrew].

Paltin, R. (1996). Makura farm. *Renewed Agriculture*, 3:26–27. [In Hebrew].

Passidomo, C. (2017). "Our" culinary heritage: Obscuring inequality by celebrating diversity in Peru and the US South. *Humanity & Society*, 41(4), 427–445.

Paxson, H. (2010). Locating value in artisan cheese: Reverse engineering terroir for new-world landscapes. *American Anthropologist*, 112(3), 444–457.

Paxson, H. (2012). *The life of cheese: Crafting food and value in America*. University of California Press.

Peck, J., & Tickell, A. (2002). Neoliberalizing space. *Antipode*, 34(3), 380–404.

Peterson, R. A. (2005). Problems in comparative research: The example of omnivorousness. *Poetics*, 33(5–6), 257–282.

Peterson, R. A., & Kern, R. M. (1996). Changing highbrow taste: From snob to omnivore. *American Sociological Review*, 900–907.

Pfeifer, G. (2019). From state-bound subjects to global subjects: Notes toward an Althusserian theory of globalized subjectivity. *Globalizations*, 1–13.

Phillips, L. (2006). Food and globalization. *Annual Review of Anthropology*, 35(1), 37–57.

Pinto, G. (2007, May 2). Pioneers of an organic lifestyle. *Haaretz*. https://www. haaretz.com/1.4821924

Pleyers, G. (2010). *Alter-globalization: Becoming actors in a global age*. Polity.

Polanyi, K. (1957). *The great transformation*. Beacon Press.

Pollan, M. (2006). *The omnivore's dilemma: A natural history of four meals*. Penguin.

Pollan, M. (2007, January 28). Unhappy meals. *New York Times Magazine*. https:// www.nytimes.com/2007/01/28/magazine/28nutritionism.t.html

Pollan, M. (2013). *Food rules: An eater's manual*. Penguin.

Pratt, A. C., & Hutton, T. A. (2013). Reconceptualising the relationship between the creative economy and the city: Learning from the financial crisis. *Cities*, 33, 86–95.

Ram, U. (1999). Between nation and corporations: Liberal post-Zionism in the global age. *Israeli Sociology*, 2(1), 99–145. [In Hebrew].

Ram, U. (2004). Glocommodification: How the global consumes the local-McDonald's in Israel. *Current Sociology*, 52(1), 11–31.

Ram, U. (2008). *The globalization of Israel: McWorld in Tel Aviv, Jihad in Jerusalem*. Routledge.

Ram, U. (2011). *Israeli nationalism: Social conflicts and the politics of knowledge*. Routledge.

Ram, U. (2016). Hebrew culture in Israel: Between Europe, the Middle East, and America. In Ben-Rafael, E., Schopes, J. H., Sternberg, Y., & Glockner, O. (Eds.), *Handbook of Israel—Major Debates* (pp. 60–75). De Gruyter.

Ran-Shachnai, I. (2015). *Loneliness and togetherness intermingled here: A spatial analysis of the Kibbutz dining hall* [MA Thesis]. Open University, Israel. [In Hebrew].

Ranta, R. (2015). Food and nationalism: From foie gras to hummus. *World Policy Journal, 32*(3), 33–40.

Ranta, R., & Mendel, Y. (2014). Consuming Palestine: Palestine and Palestinians in Israeli food culture. *Ethnicities, 14*(3), 412–435.

Raviv, Y. (2003). Falafel: A national icon. *Gastronomica: The Journal of Food and Culture, 3*(3), 20–25.

Raviv, Y. (2015). *Falafel nation: Cuisine and the making of national identity in Israel.* University of Nebraska Press.

Ray, K. (2016). *The ethnic restaurateur.* Bloomsbury.

Raynolds, L. T. (2000). Re-embedding global agriculture: The international organic and fair-trade movements. *Agriculture and Human Values, 17*(3), 297–309.

Raynolds, L. T. (2004). The globalization of organic agro-food networks. *World development, 32*(5), 725–743.

Redden, G. (2011). Religion, cultural studies and New Age sacralization of everyday life. *European Journal of Cultural Studies, 14*(6), 649–663.

Reed, M. (2001). Fight the future! How the contemporary campaigns of the UK organic movement have arisen from their composting of the past. *Sociologia Ruralis, 41*(1), 131–145.

Regev, M. (2007). Cultural uniqueness and aesthetic cosmopolitanism. *European Journal of Social Theory, 10*(1), 123–138.

Regev, M., & Seroussi, E. (2004). *Popular music and national culture in Israel.* University of California Press.

Renting, H., Marsden, T. K., & Banks, J. (2003). Understanding alternative food networks: Exploring the role of short food supply chains in rural development. *Environment and Planning A, 35*(3), 393–411.

Reynolds, K. (2015). Palestinian agriculture and the Israeli separation barrier: The mismatch of biopolitics and chronopolitics with the environment and human survival. *International Journal of Environmental Studies, 72*(2), 237–255.

Ritzer, G. (2004). *The McDonaldization of society.* Pine Forge Press.

Ritzer, G. (2007). *The globalization of nothing 2.* Pine Forge Press.

Robertson, R. (1995). Glocalization: Time-space and homogeneity-heterogeneity. *Theory, Culture & Society, 36*(1), 25–45.

Robinson, J., & Farmer, J. (2017). *Selling local: Why local food movements matter.* Indiana University Press.

Roniger, L., & Feige, M. (1992). From pioneer to freier: The changing models of generalized exchange in Israel. *European Journal of Sociology, 33*(2), 280–307.

Rosa, H. (2004). Four levels of self-interpretation: A paradigm for interpretive social philosophy and political criticism. *Philosophy & Social Criticism, 30*(5–6), 691–720.

Rosenthal, G., & Eiges, H. (2014). Agricultural cooperatives in Israel. *Journal of Rural Cooperation, 42*(8), 1–29.

Rosner, M. (2000). Future trends of the Kibbutz: An assessment of recent changes. Publication no. 83. Institute for Study and Research of the Kibbutz, Haifa University. http://research.haifa.ac.il/~kibbutz/pdf/trends.PDF

Rothenberg, J., & Fine, G. A. (2008). Art worlds and their ethnographers. *Ethnologie Française*, 38(1), 31–37.

Rozin, O. (2006). Food, identity, and nation-building in Israel's formative years. *Israel Studies Review*, 21(1), 52–80.

Ruah-Midbar, M., & Zaidman, N. (2013). "Everything starts within": New Age values, images, and language in Israeli advertising. *Journal of Contemporary Religion*, 28(3), 421–436.

Rudy, K. (2012). Locavores, feminism, and the question of meat. *The Journal of American Culture*, 35(1), 26–36.

Russell, R., Hanneman, R., & Getz, S. (2011). The transformation of the kibbutzim. *Israel Studies*, 16(2), 109–126.

Sade, S. (2011, October 16). Tnuva—From "Israeli brand" to "public enemy." *TheMarker*. http://archive.is/kuXpz#selection-1435.108-1435.198 [In Hebrew].

Sasson, T. (2005, March 10). *Opinion concerning unauthorized outposts*. Prime Minister's Office, Israel. www.mfa.gov.il/mfa/aboutisrael/state/law/pages/summary%20of%20opinion%20concerning%20unauthorized%20 outposts%20-%20talya%20sason%20adv.aspx

Sayre, L. (2011). The politics of organic farming: Populists, evangelicals, and the agriculture of the middle. *Gastronomica: The Journal of Food and Culture*, 11(2), 38–47.

Schiermer, B. (2014). Late-modern hipsters: New tendencies in popular culture. *Acta Sociologica*, 57(2), 167–181.

Schlosser, E. (2001). *Fast food nation: The dark side of the all-American meal*. Houghton Mifflin.

Schnell, I., & Mishal, S. (2005). *Uprooting and settlers' discourse: The case of Gush Katif*. The Floersheimer Institute for Policy Studies. http://upload-community.kipa.co.il/52920123662.pdf [In Hebrew].

Schudson, M. (1989). How culture works. *Theory and Society*, 18(2), 153–180.

Schwartz, M. (1995). *Unlimited guarantees: History, political economy and the crisis of cooperative agriculture in Israel*. Ben Gurion University of the Negev Press. [In Hebrew].

Schwarz, O. (2020). Identity as a barrier: claiming universality as a strategy in the Israeli vegan movement. *Social Movement Studies*, 1–19.

Scott, M. (2017). 'Hipster capitalism' in the age of austerity? Polanyi meets Bourdieu's new petite bourgeoisie. *Cultural Sociology*, 11(1), 60–76.

Scrinis, G. (2008). On the ideology of nutritionism. *Gastronomica*, 8(1), 39–48.

Scrinis, G. (2013). *Nutritionism: The science and politics of dietary advice*. Columbia University Press.

Sela, R. (2012, January 1). Natural phenomena: The story of Eden Teva Market. NRG. http://www.nrg.co.il/online/21/ART2/325/745.html [In Hebrew].

Senor, D., & Singer, S. (2009). *Start-up nation: The story of Israel's economic miracle*. Twelve.

Sewell, W. H. Jr. (1992). A theory of structure: Duality, agency, and transformation. *American Journal of Sociology*, 98(1), 1–29.

Shafir, G. (1996). *Land, labor and the origins of the Israeli-Palestinian conflict, 1882–1914*. University of California Press.

Shafir, G. (2017). *A half century of occupation: Israel, Palestine, and the world's most intractable conflict*. University of California Press.

Shafir, G., & Peled, Y. (2002). *Being Israeli: The dynamics of multiple citizenship*. Cambridge University Press.

Shaili, G. (2006, June 28). The Israeli organic. *Yedioth Ahronoth*. [In Hebrew].

Shalev, M. (2000). Liberalization and the transformation of the political economy. In Shafir, G., and & Peled, Y. (Eds.), *The new Israel: Peacemaking and liberalization* (pp. 129–159). Routledge.

Shalev, M. (2001, January 12). The hummus is ours. *Yediyout Aharonot*. [In Hebrew].

Shani, L. (2018). Of trees and people: The changing entanglement in the Israeli desert. *Ethnos*, 83(4), 624–644.

Shapin, S. (2006). Paradise sold. *New Yorker*, 15, 84–88.

Shapira, A. (2004). The Bible and Israeli identity. *AJS review*, 28(1), 11–41.

Shapira, A. (2012). *Israel: A history*. Brandeis University Press.

Shenhav, Y. (2007). Modernity and the hybridization of nationalism and religion: Zionism and the Jews of the Middle East as a heuristic case. *Theory and Society*, 36(1), 1–30.

Shoham, H. (2017). *Israel celebrates: Jewish holidays and civic culture in Israel*. Brill.

Shragai, N. (2007, August 1). Good eggs from the West Bank. *Haaretz*. https://www.haaretz.com/1.4957720

Shuk Hanamal. (n.d.). About. https://shukhanamal.co.il/about/

Simchai, D. (2009). *Flowing against the flow: Paradoxes in realizing New Age vision in Israel*. Pardes. [In Hebrew].

Simchai, D., & Keshet, Y. (2016). New Age in Israel: Formative ethos, identity blindness, and implications for healthcare. *Health*, 20(6), 635–652.

Simmel, G. (1997). Sociology of the meal. In Frisby, D., & Featherstone, M. (Eds.), *Simmel on culture: Selected writings* (pp. 130–135). Sage. (Original work published 1915)

Simovitch, O. (2015). *"Back to the sources" through imported models: Transformations in the value of olive oil in Israel* [MA thesis]. The Open University of Israel. [In Hebrew].

Sklair, L. (2001). *The transnational capitalist class*. Blackwell.

Smilansky, M. (1926). *The agricultural settlement*. Shahar Press. [In Hebrew].

228 REFERENCES

Soffer, O. (2015). *Mass communication in Israel: Nationalism, globalization, and segmentation*. Berghahn Books.

Sorek, B. (2008). *Bringing forth bread from the earth, together—A story of community supported agriculture in Israel* [MA thesis]. New College of California.

Sorek, B. (2012). Aley Chubeza #112—May 21st–23rd 2012—Shavuot. Chubeza. http://chubeza.com/?p=6274&lang=en

Stephenson, G. (2008). *Farmers' markets: Success, failure, and management ecology*. Cambria Press.

Sternhell, Z. (1998). *The founding myths of Israel: Nationalism, socialism and the making of the Jewish state*. Princeton University Press.

Swedenburg, T. (1990). The Palestinian peasant as national signifier. *Anthropological Quarterly, 63*(1), 18–30.

Swidler, A. (1986). Culture in action: Symbols and strategies. *American Sociological Review, 51*(2), 273–286.

Szabo, M., & Koch, S. (Eds.) (2017). *Food, masculinities, and home: Interdisciplinary perspectives*. Bloomsbury.

Szasz, A. (2007). *Shopping our way to safety: How we changed from protecting the environment to protecting ourselves*. University of Minnesota Press.

Tal, A. (2007). To make a desert bloom: The Israeli agricultural adventure and the quest for sustainability. *Agricultural History, 81*(2), 228–257.

Talshir, R. (2012). *The secret method: 100 simple rules for healthy eating*. Keter. [In Hebrew].

Tavory, I., & Goodman, Y. C. (2009). "A collective of individuals": Between self and solidarity in a Rainbow gathering. *Sociology of Religion, 70*(3), 262–284.

Taylor, C. (1992). *The ethics of authenticity*. Harvard University Press.

Tesdell, O. I. (2015). Territoriality and the technics of drylands science in Palestine and North America. *International Journal of Middle East Studies, 47*(3), 570–573.

Thompson, C. J., & Coskuner-Balli, G. (2007). Countervailing market responses to corporate co-optation and the ideological recruitment of consumption communities. *Journal of Consumer Research, 34*(2), 135–152.

Tiemann, T. (2008). Grower-only farmers' markets: Public spaces and third places. *Journal of Popular Culture, 41*(3), 467–487.

Tnuva. (n.d.). About Harduf. http://en.tnuva.co.il/products/5

Tomlinson, J. (1999). *Globalization and culture*. University of Chicago Press.

Tovey, H. (1999). 'Messers, visionaries and organobureaucrats': Dilemmas of institutionalisation in the Irish organic farming movement. *Irish Journal of Sociology, 9*(1), 31–59.

Tovey, H. (2009). 'Local food' as a contested concept: Networks, knowledge and power in food-based strategies for rural development. *International Journal of Sociology of Agriculture and Food, 16*(2), pp. 21–35.

Trivette, S. A. (2019). The importance of food retailers: applying network analysis techniques to the study of local food systems. *Agriculture and Human Values, 36*(1), 77–90.

Trubek, A. (2008). *The taste of place: A cultural journey into terroir*. University of California Press.

Trubek, A., Guy, K. M., & Bowen, S. (2010). Terroir: A French conversation with a transnational future. *Contemporary French and Francophone Studies*, *14*(2), 139–148.

Tzahor, Z. (2010). Others. In Halamish, A., & Tzameret, Z. (Eds.), *The Kibbutz: The first 100 Years* (pp. 379–390). Yad Ben-Zvi. [In Hebrew].

Tzfadia, E. (2008). Geography and demography: Spatial transformations. In Ben-Porat, G., Levy, Y., Mizrahi, S., Naor, A., & Tzfadia, E. (Eds.), *Israel since 1980* (pp. 42–68). Cambridge University Press.

Tzfadia, E. (2008b). Abusing multiculturalism: The politics of recognition and land allocation in Israel. *Environment and Planning D: Society and Space*, *26*(6), 1115–1130.

T'zvi, L. (2010, 18 November). Local consumption and environmental discourse. *Saloona*. http://saloona.co.il/zviliat/?p=70?ref=asur [In Hebrew].

Urciuoli, B. (2008). Skills and selves in the new workplace. *American Ethnologist*, *35*(2), 211–228.

Uriely, N., Yonay, Y., & Simchai, D. (2002). Backpacking experiences: A type and form analysis. *Annals of Tourism Research*, *29*(2), 520–538.

Uzzi, B. (1996). The sources and consequences of embeddedness for the economic performance of organizations: The network effect. *American Sociological Review*, *61*(4), 674–698.

Veblen, T. (1934). *The theory of the leisure class: An economic study of institutions*. The Modern Library. (Original work published 1899)

Ventura, D. (n.d.). Har-Sinai farm. Ventura Daniel Wiki. https://danielventura. wikia.org/he/wiki/חוות_הר_סיני_סוסיא [In Hebrew].

Vered, R. (2010, May 14). Working clean. *Haaretz*. [In Hebrew].

Vered, R. (2012, May 23). The vegetable boutique. *Haaretz*. https://www.haaretz. co.il/food/dining/1.1714340 [In Hebrew].

Vered, R. (2013, January 9). Fantastic peas at the foot of the mountain. *Haaretz*. https://www.haaretz.co.il/food/dining/1.1903824 [In Hebrew].

Vered, R. (2015, December 29). Is there anything to say about hummus that has not been said so far? *Haartez*. https://www.haaretz.co.il/food/dining/. premium-1.2810117 [In Hebrew].

Vered, R., Cohen, H., & Landau, E. (2010, September 17). The organic cock-crow. *Haaretz*. [In Hebrew].

Verlinsky N. (1960). *Tnuva: The cooperative marketing society for cooperative produce*. Afro-Asian Institute Publications.

Verter, B. (2003). Spiritual capital: Theorizing religion with Bourdieu against Bourdieu. *Sociological Theory*, *21*(2), 150–174.

Voget, G. (2007). The origins of organic farming. In Lockeretz, W. (Ed.) *Organic farming: An international history* (pp. 9–29). CABI.

Volkomir, K. (2010, October 18). Urban plots in the market. *Check.Eat.Out*. http://www.checkeatout.co.il/?cat=10 [In Hebrew].

Vos, T. (2000). Visions of the middle landscape: Organic farming and the politics of nature. *Agriculture and Human Values, 17*(3), 245–256.

Warde, A. (2016). *The practice of eating.* Polity.

Warde, A., Wright, D., & Gayo-Cal, M. (2007). Understanding cultural omnivorousness: Or, the myth of the cultural omnivore. *Cultural sociology, 1*(2), 143–164.

Watson, J. (2006). *Golden arches east: McDonald's in East Asia* (2nd ed.). Stanford University Press.

Watson, J., & Caldwell, M. L. (Eds.) (2005). *The cultural politics of food and eating: A reader.* Blackwell.

Watson, S. (2009). The magic of the marketplace: Sociality in a neglected public space. *Urban Studies, 46*(8), 1577–1591.

Weiss, E. (2016). 'There are no chickens in suicide vests': The decoupling of human rights and animal rights in Israel. *Journal of the Royal Anthropological Institute, 22*(3), 688–706.

Werczberger, R., & Huss, B. (2014). New age culture in Israel. *Israel Studies Review, 29*(2), 1–16.

Wexler, A. (2016). Seeding controversy: Did Israel invent the cherry tomato?. *Gastronomica: The Journal of Critical Food Studies, 16*(2), 1–11.

Wilk, R. (2006). *Home cooking in the global village: Caribbean food from buccaneers to ecotourists.* Berg.

Wilkins, J. L. (2005). Eating right here: Moving from consumer to food citizen. *Agriculture and Human Values, 22*(3), 269–273.

Williams, G. (2010). *The knowledge economy, language and culture.* Multilingual Matters.

Wolinitz, D. (1998). The state of Israel: Towards the world's first organic state. *Organic, the IBOAA Bulletin 3:3.* [In Hebrew].

Wolinitz, D. (1999). The dream of the organic state: Towards the realization of the dream. *Organic, the IBOAA Bulletin 4:3.* [In Hebrew].

Wood, M. (2007). *Power, possession and the New Age: Ambiguities of authority in neo-liberal societies.* Ashgate.

Wood, M., & Bunn, C. (2009). Strategy in a religious network: A Bourdieuian critique of the sociology of spirituality. *Sociology, 43*(2), 286–303.

Yates, L., & Warde, A. (2017). Eating together and eating alone: meal arrangements in British households. *The British Journal of Sociology, 68*(1), 97–118.

Yefet, O. (2009, October 15) The founder and CEO of Eden Teva Market: "Tnuva knows that I will succeed." *Calcalist.* https://www.calcalist.co.il/marketing/articles/0,7340,L-3364108,00.html [In Hebrew].

Yurista, D. (2015, May 27). Organic agriculture in Israel [Press announcement]. Ministry of Agriculture and Rural Development, Israel (MARD). https://www.moag.gov.il/yhidotmisrad/dovrut/publication/2015/Pages/organic_fruits_veg.aspx [In Hebrew].

Zaban, H. (Ed.). (2012). *150 Years of agriculture in Israel*. Maariv Publication. [In Hebrew].

Zali, R. (1993). From the secretariat. *Renewed Agriculture*, 23:14. [In Hebrew].

Zelizer, V. (1985). *Pricing the priceless child: The changing social value of children*. Basic Books.

Zelizer, V. (2011). *Economic lives: How culture shapes the economy*. Princeton University Press.

Zerubavel, Y. (1996). The Forest as a national icon: Literature, politics and the archeology of memory. *Israel Studies*, *1*(1), 60–97.

Zerubavel, Y. (2008). Memory, the rebirth of the native, and the "Hebrew Bedouin" identity. *Social Research: An International Quarterly*, *75*(1), 315–352.

Zimmerman, C. C. (1928). The family budget as a tool for sociological analysis. *American Journal of Sociology*, *33*(6), 901–911.

Zukin, S. (2008). Consuming authenticity: From outposts of difference to means of exclusion. *Cultural Studies*, *22*(5), 724–748.

Zukin, S., & DiMaggio, P. (Eds.). (1990). *Structures of capital: The social organization of the economy*. Cambridge University Press.

# Index

Note: Page numbers in *italics* indicate illustrations.